They Gave Blood, Sweat, And Tears

True Stories of American Veterans From World War II to Operation Iraqi Freedom

By

Bill Hall

Other books by Bill Hall

Turnout-A Firefighters Story

Turnout-Second Alarm

Sykesville

Remember Me in Paradise-The Story of the Thief

ISBN: 978-1-62249-589-4

Published by
Biblio Publishing
Columbus, Ohio
BiblioPublishing.com

DEDICATION

This book is dedicated to the millions of American veterans who have served in the American Armed Forces. Their dedication and fortitude in the presence of enemy forces, seen and unseen, and during peacetime, is unwavering and is to be commended daily. The oath they swear when entering the service is one they take not lightly:

"I, _____, do solemnly swear that I will support and defend the Constitution of the United States against all enemies, foreign and domestic; that I will bear true faith and allegiance to the same; and that I will obey the orders of the President of the United States and the orders of the officers appointed over me, according to regulations and the Uniform Code of Military Justice. So help me God."

It is with a grateful heart that I give thanks to God for allowing me the opportunity to interview so many American heroes and to hear and tell their stories. This book is dedicated to them, their heroism, their sacrifice and their service. There but by the Grace of God they go.

CONTENTS

They Gave Blood, Sweat and Tears

Introduction

The stories you are about to read are real. I was able to interview fifty-eight veterans during the course of thirteen months from 2009-2010 and tell their stories to the public. Fifty-four have agreed to let you hear them. Let me tell you how this project began.

In the early spring of 2009, I was between jobs. One day I decided to go out for lunch and visited a local fast food restaurant in the area. As I sat there and ate my lunch, I noticed a car pull up outside. From the vehicle an elderly gentleman slowly got out and went to the passenger side and assisted his wife from the car. She was obviously blind and needed assistance.

The pair cautiously entered the restaurant, and he guided his wife into her seat. He then went to the counter and ordered their food. As I glanced over from time to time, I watched them eat and converse happily with one another.

I noticed the ball cap the man was wearing but could not read it very well from where I was sitting. I did realize it had something to do with being a veteran. I got up to get a refill of my drink and walked past them. I was able to read his hat at that point. It said that he had been in World War II and had been in an amphibious group in the Navy.

I went back to my seat and finished my lunch. I pondered my next move. I gathered up my trash and got up and put it in the garbage container. I then walked over to the gentleman. I asked him if he had been in World War II and he acknowledged that he had been and served in the Navy. I reached out with my right hand and shook his, thanking him for his service to our country. Delighted and overwhelmed, he tried to stand up but I begged him to stay. "Why thank you. Were you in the service, son?" He inquired of me.

"Yes." I replied. "I was in the Coast Guard during Vietnam." He then reciprocated and thanked me for my service. I smiled at both of them and made my way out to my car. When I got there, I sat for a while and did some thinking.

Someone needs to tell these stories, I thought to myself, *before it's too late*. As I began the drive home, my mind would not stop thinking about this gentleman. By the time I arrived home, I had a plan thought out.

I got on my computer and created a query on email to send to the Carroll County Times newspaper in Westminster, Maryland. It would be to the attention of the editor, Mr. Jim Lee. I asked Mr. Lee if he would consider a newspaper series on true stories of veterans. I would go out and interview each one and then tell their story. He thought it was a great idea. He requested five to six stories ahead to see how they would look and see my writing style. The rest is history. From July 4, 2009 until August 7, 2010 the series *Blood, Sweat and Tears* appeared weekly in the Times.

I never did get the name of that gentleman in the fast-food restaurant that fateful day. But I am indebted to him and to every veteran I interviewed in the series. I have never felt as humbled as I did when I did each interview in the presence of true American heroes.

One thing that really amazed me was how humble these gentlemen were. Not one of them glorified war. They all said they were just doing their job for their country. They did not consider themselves special. They did not consider themselves heroes. They did not think they deserved special treatment because of what they did. They did it because they felt it was their duty.

"*War is Hell*." These were the words of General William Tecumseh Sherman during the American Civil War. These words would resonate throughout the decades to follow as an apt description of armed conflict.

Many people try to glorify and make war seem to be something to revere. In their eyes there is nothing like it. There is nothing ever glorious from war, not even the victory. Lives are lost, bodies maimed for life, civilians killed, and properties are destroyed. The joy of the victories may be short-lived due to the overwhelming realization that a country must be rebuilt and families must be reborn.

What the men in these stories did was an ultimate test of their character and fortitude. Their character speaks volumes as they reached heights never approached before. Many were from farms or small towns. Some withdrew from college and put their education on hold. They all left families and sweethearts behind. They knew what they were going to was much larger and important. What they were going to do had to be done and they would do it. Some came home perfectly fine while others took years to regroup, and some would never be able to. Some were prisoners-of-war and some may never have seen battle. But all had served and did so without reservation.

These gentlemen are the real heroes in American history and everyday life, not television or movie stars, rockers or sports people. They are the ones that have helped preserve our liberties and the lives we have today. They are the ones that have lived and died for our freedoms abroad, so we did not have to fight them on our ground. Wars are not something we should strive for. They are something that should be avoided at all costs. However, when all efforts fail, it may be a necessary evil. If we are to preserve our way of life, it comes with a cost. That cost may be war and loss of lives.

There is a saying that nothing is free. It is so true when it comes to freedom. Our nation has been truly blessed by God and He has given us the rights and freedoms to preserve it and maintain it. We must be good stewards of what He has given unto us.

Decades ago, the military used to use a phrase that is still very appropriate – peace through strength. If we do not maintain what these men have fought for, it will be lost. We must maintain a strong national defense and preserve the liberties of our nation for our future generations. There will always be evil in the world and we must be ready to face it. We must be strong enough to face. We must have the guts to face it. When and if war comes to us, we must be ready to fight it to win. We must have men and women with the moral fortitude and courage to face it head on such as those in these stories.

I publicly thank these men for allowing me to tell their stories to the world. During these interviews sometimes chins quivered, sometimes eyes welled up, and sometimes exits from the room were appropriate. But the stories were told. They felt equally compelled to let people know what they did and what they saw. People should know what they sacrificed in time, family, friends and memories.

In closing, please enjoy these stories and when you see a veteran or know a veteran, take a moment to say "Thank you." Look around and thank God for the nation you live in and the freedoms you enjoy. It is because of them and **not** a politician you have the liberties you have today.

A special thank you goes to Mr. Jim Lee of the Carroll County Times newspaper for having the faith in me to allow me to tell these stories in his newspaper. I would also like to thank Mr. Wayne Carter of the Carroll County Times for working with me weekly in getting the stories edited, corrected and published in a timely manner.

Bill Hall

Eliot W. Annable
Aka Bill Eliot
Master Sergeant, U.S. Army
423rd Regimental Headquarters Co.
106th Infantry Division
World War II

Little did Charles and Elizabeth Annable of Bangor, Maine know that on October 13, 1924, their newborn son, Eliot, would one day become a decorated warrior in one of the greatest wars of the ages.

Just eighteen years old and in his first year at the University of Rhode Island, Eliot was inducted into the Army on June 11, 1943. During his service as a combat infantryman, he received combat credit (battle stars on his European campaign ribbon) for Northern France, Central Europe, the Ardennes (Battle of the Bulge), and the Rhineland. He was later awarded the Bronze Star for his combat service.

On Dec. 16, 1944 the 106th Infantry Division, consisting of the 422nd, 423rd and 424th Infantry Regiments and supporting elements, was spread thinly just inside the western border of Germany facing the heavily fortified German Siegfried Line. Eliot's company, the 423rd Regimental Headquarters Co., occupied Buchet, Germany, a small hamlet just east of St. Vith, Belgium. The whole area was considered to be a quiet sector and it was felt that the Germans were not capable of mounting a significant attack in that area. As a complete surprise to the allied leaders, at 5:30 AM, on that Dec. 16th the Battle of the Bulge began with a massive ground shaking German artillery barrage followed by an overwhelming attack by armored panzer divisions and infantry. By the end of the second day the

422nd and 423rd Regiments were surrounded with no chance to withdraw or to receive any reinforcing support. Although almost out of ammunition and completely cut off, the 422nd and 423rd Regiments held out for a week before being forced to surrender. Less than a month later Eliot would be the only remaining member of his platoon not killed, wounded, or captured.

When the battle started Eliot was a radio operator in the company's communications platoon with the rank of corporal. Immediately after the attack started communication between units was almost exclusively by radio since most of the wires had been severed. In fact, the first casualty in Eliot's platoon was a wireman who was killed by German fire while on a pole trying to repair severed wires. Because of these communication problems, Eliot and Herb Heidepriem, another radio operator, were sent with a jeep mounted radio, and with armored car escort, to an attached cavalry troop, Troop B 18th Cavalry, to establish communications. Troop B was in Winterscheid, Germany, a small town a few miles away from 423rd headquarters. Troop B had been under heavy attack and by nightfall of the first day the Germans had driven a wedge between Troop B and the 423rd and had penetrated deeper to the rear of the American lines as they were concentrating on their first major objective, the city of St. Vith, Belgium.

Early on the second day Troop B was ordered to attempt to rejoin the regiment in a defense perimeter. At Mutzenich, Germany, Troop B was blocked from getting back to the 423rd by columns of German forces on one of the main roads to St. Vith. After heavy artillery shelling a German under a white flag demanded that Troop B surrender. The demand was refused, and Eliot reported to regiment that they were waiting for further orders. When none came, Captain Fossland, the commander of Troop B, decided to attempt to break through to St. Vith several miles to the west. This attempt was blocked by a convoy of German troops and after a brief firefight, and with all escape routes blocked, Troop B withdrew to the village of Amelscheid to take stock. Finally establishing radio contact with the 423rd Captain Fossland was told "St. Vith is in enemy hands. Make your own decision." Captain Fossland then ordered the troop to withdraw into a densely wooded area, destroy as much equipment as they could, separate into small groups, and attempt to infiltrate back to American forces west of St. Vith. St. Vith was actually still in American hands and held out for four more days causing a major setback to the Germans.

Eliot and Herb destroyed as much of their radio equipment as they could, pulled the wires from their Jeep and headed into the Ardennes Forest. It was foggy, the ground was covered with patchy snow, and the sky was completely overcast. They had no food, maps, or compass. To protect themselves they had their M1 rifles and 40 rounds of ammunition. For the next five days, although fired on several times, they managed to avoid capture by keeping to the woods and skirting towns as much as possible. At night, with no visibility to stars, they used the glow of fires from burning towns to help in heading west. They were also aided by the sounds of German V1 missile "buzz bombs" being fired towards Liege, Belgium. At one point, while hiding along the side of a road, they heard a convoy of American vehicles approaching. They kept hidden and saw, as the trucks passed by, they were captured American trucks filled with German soldiers. Eliot later learned the road led to Malmady where the infamous Malmady massacre of American POW's occurred that day.

After five days of working their way west, Eliot and Herb finally encountered elements of the 82nd Airborne Division near Trois Ponts, Belgium. They were held at rifle point until they could prove they were Americans because German troops had been infiltrating dressed in American uniforms. They were returned to the only remaining regiment of the 106th Division, the 424th. For the rest of the war in

Europe the 424th operated as a regimental combat team filling gaps between other divisions.

On January 13, 1945 while elements of the 424th were attempting to take the town of Coulee, Belgium, Herb was severely wounded by artillery fire and was hospitalized for the duration. That made Eliot the only member of his original platoon still in action (Herb later became one of the leading legislators in South Dakota). Eliot continued with the 424th for the rest of the war. Soon after Germany surrendered, he was transferred to the 3rd Armored Division, where he was promoted to Master Sergeant as regimental communications chief, and then to the 2nd Armored Division for return to the US and discharge on Feb. 6, 1946.

For his service, Eliot was awarded five battle stars, the Combat Infantry Badge, Expert Rifle, European African Middle Eastern Theater Campaign Ribbon, Good Conduct Medal and the Bronze Star. To him, the most meaningful is the Bronze Star, an honor more for the men who didn't survive than for himself.

Eliot moved to Maryland in 1949 after finishing college and worked for Westinghouse as an electronics engineer for 36 years until 1984. Today he resides in New Windsor, Maryland and works as a musician in the Baltimore Musicians Union. In regard to his service, he says "It was an honor to serve my country, and I would do it again under similar circumstances, but there is nothing glorious about war and its horrors. It is all misery and suffering. I hope the day comes when all countries will learn that there are no winners in a war… only losers."

When he returned home, he recalls sitting in his bedroom thinking as he reached out and touched the walls, "I am really here. I can't believe I made it!"

James F. Bard, Jr.
Captain
Jet Mechanic/Comptroller
91st Strategic Reconnaissance Wing
U.S. Air Force
Cold War

Beale Air Force Base, California…January 1966. Many of the personnel stationed at Beale, including twenty-nine-year-old James Bard, Jr., ran out onto the runway to see what was coming in for a landing. They lined the tarmac and waited patiently. Then out of the blue, a glint appeared. Slowly, it began to develop a shape that became recognizable. It began its approach and landed quietly. Bard never saw anything so beautiful in his life. The Lockheed SR-71 long range strategic reconnaissance aircraft, unofficially named "Blackbird," rolled to a stop from its birthplace in Sunnyvale, California. He could not stop looking at it. It was a moment he would never forget.

James F. Bard, Jr. was born February 19, 1937 to Mary Catherine and James Bard, Sr. of Hagerstown, Maryland and lived there with his brother and sister. In 1942, his father took the family and moved to Baltimore, Maryland to work at Maryland Dry Dock. Bard, Jr. attended high school at Thomas Edison but quit at age 16. He went to work for Maryland Casualty Insurance Company for a year as an IBM machine operator. Duty called, but Mom had to sign, and James Bard enlisted in the U.S. Air Force at age 17 on March 29, 1954.

Bard wound up at Sampson Air Force Base in New York for recruit training. His first assignment was Flight 3244 but lasted for only a week due to him receiving pneumonia. He says he left the meanest instructor for the greatest instructor. The young recruit wound up at Amarillo AF Base, Texas to be trained as a multi-jet aircraft mechanic for the B-47.

In November 1954, he received his first permanent assignment to the 323rd SR Squadron of the 91st Strategic Reconnaissance Wing at Lockbourne Air Force Base in Ohio as an RB-47E mechanic. While there he met his first wife at a USO dance, married her in 1957 and had three children before divorcing in 1968. He began looking for an inside job and, because of his experience in IBM machines, began working in the office of the Chief of Maintenance on the new time accounting system recently instituted throughout the Strategic Air Command.

The 91st has a long history. It was the 91st Bomb Group during WW II but was deactivated after the war. It was then reactivated as the Strategic Reconnaissance Wing under the Strategic Air Command and during Vietnam it became a wing for B-52's. In 1968 it would become the 91st Missile wing.

In the summer of 1956, the wing went to Ben Guerir AB in Morocco. However, in November 1957 the 91st was deactivated and were re-designated the 376th Bombardment Wing. They worked on EB-47E's which were full of electronic countermeasures equipment. But in 1958, his enlistment ended, and he left the Air Force for 85 days but realized he made more money in the service as a Staff Sergeant. So he reenlisted.

With the new Maintenance Data Collection System instituted, he found himself implementing the new system for the 376th Bomb Wing. He attended specialized training at Chanute AFB in Illinois and then went to Oklahoma City, Oklahoma to assist in writing the system codebook for the B-47's.

In 1962 he changed career paths to get out of maintenance. He wound up being reassigned from aircraft maintenance to the Management Analysis Office in Wing

Headquarters. Bard studied for and passed his GED and enrolled in evening college classes and in December 1963 was promoted to Technical Sergeant and sent to Goose Bay AB in Labrador, Canada. Fifteen months later he was assigned to Beale AFB, California with the SR-71 Blackbird Program, later designed the 9th Strategic reconnaissance Wing, receiving promotion to Master Sergeant in 1966.

While at Beale, he took more college courses and graduated from Yuba Community College and began attending the University of Nebraska at Omaha in 1968. While there, he was accepted into Officers Training School (OTS). In May 1969, he received his Bachelor's and headed off to OTS at Lackland AFB, Texas, graduating as a Second Lieutenant September 6, 1969. Bard was assigned to the Air Force Office of Special Investigations in Washington, DC where he received his next two promotions. He also received another promotion in 1971 when he married Ruth, his second wife of 39 years, in Omaha.

In August 1973, Bard found himself assigned to Hahn, Germany for the next four years as Chief of Management and Budget for the 50th Tactical Fighter Wing. In 1977 assignment came to the 347th Tactical Fighter Wing at Moody AFB, Georgia where he would be Base Comptroller for three months. But on October 1, 1980, Bard had set his sights on retirement. He is proud of rising in the ranks from Basic Airman to Captain. He talks about how proud he is to have served his country and improve his own life over and over.

During the Cold War, pilots were shot down and aircraft were lost. Aerial surveillance and covert flights was a common thing on both sides. Bard tells tales of pilots coming back from missions with their places full of holes from MIG-17's.

Bard and his family settled down in Carroll County in 1981. He loves to tell tales of his service to his country so much that he has co-authored a book with Linda Rios

Bromley and Frank T. Hayes titled "We Served with Honor-Memories of the Men who served the 91st Strategic Reconnaissance Wing" published by McDowell Publishing in 2009. Today he enjoys his three children and four stepchildren, six grandchildren and two great grandchildren. He also attends annual reunions that began in 1996. Today he is the Secretary of the 91st Strategic Reconnaissance Wing Association.

Jerry F. Barnes
Sergeant, Green Berets
1st Special Forces Regiment,
5th Special Forces
U.S. Army
Vietnam

Twenty-one-year-old Jerry Barnes sat in the jungle with twelve Vietnamese Mountain Yards. They were from the indigenous mountain tribes of Vietnam. He was their Special Forces advisor at a Civilian Irregular Defense Group (CIDG) radio relay site. As he sat there admiring the peaceful sunset the site began receiving a heavy volume of automatic weapons fire from the west at a distance of sixty meters. That was when all Hell broke loose.

Jerry Barnes was born to Charles and Mary Barnes of Westminster, Maryland on October 9, 1948. While growing up his Dad would tell him of serving in the Army during World War II in the South Pacific with the 38th Infantry Division. He would go on to graduate from the old Westminster High School and go on to drafting school. He worked five months for a Field Engineering firm when he received his letter from Uncle Sam.

Barnes was drafted in the Army in June 1968, processed at Fort Holabird, Maryland and off to Fort Bragg, North Carolina where he received his basic training. By signing on for a third year, he was allowed to join the new Green Berets. He received radio operator training at Fort Jackson, South Carolina and then went to Fort Benning, Georgia for one month of Jump School. This was followed by two months of Green Beret training back at Fort Bragg and Special Forces Radio Operator School for two months.

After graduation they were put into A-Teams and spent two weeks in a national forest before final graduation. Finally, he was assigned to a Special Forces group, received his patch for his beret and went home for one month. He then volunteered to go to Vietnam for a one-year tour. As he left, he wondered if he would ever come back.

The young Green Beret arrived in Cam Ranh Bay, Vietnam in September 1969. He went from forty-five degrees temperature when he left home to one hundred degrees upon arrival. His first night in the Southeast Asian country was met with a rocket attack. He was sent to Na Trang for assignment and was put in an “A” Camp along the Cambodian/Laotian border just 25 miles from the Ho Chi Minh Trail to monitor enemy activities.

The Special Forces built large outposts (“A” Camps) that contained about 12 Americans and 200-300 Vietnamese soldiers with their families, many of whom were Vietnamese Special Forces. Their missions were to go out every two weeks and patrol with about 50-100 Mountain Yards to locate Viet Cong.

Each “A” Camp was assigned specialties. Barnes was a trained radio operator and small weapons specialist. They also had specialists in weapons, engineering and demolition, intelligence, and medical. There were usually two officers, a Lieutenant was the XO and a captain was the team commander. Most of the soldiers in the Special Forces were career people. When they began taking heavy losses, they started recruiting three-year soldiers.

Barnes was amazed at the wildlife. Tigers could be seen, and one even chased a company of men and they had to call in an air strike on it! Elephants, spiders and poisonous snakes were plentiful. The men had to sleep in hammocks tied to trees two feet off the ground. Early one morning a cook began screaming for help. Barnes rushed into the kitchen to see a Cobra in the sitting position ready to strike the cook.

He says he killed it but blew half the kitchen apart. The snake was killed in action.

One evening while watching the sunset, Barnes and the 12 Mountain Yards with him came under heavy automatic weapons fire from the west. He was an advisor to the group at the CIDG radio relay site. Barnes, in an effort to better control his defenses, immediately moved from his command post to the western part of the perimeter with a radio. He began directing light machine gun fire in countering the enemy.

When the enemy later opened fire on his troops from the south, he moved across the perimeter under heavy enemy fire in order to gain an accurate picture of the situation. Once there, he came across two Yards who had been wounded by small arms fire. He immediately had these two men moved to his command bunker and began moving from position to position, encouraging the Yards to remain calm and place effective fire on the enemy. Through his efforts, the enemy was repelled, and two lives were saved. Barnes would receive his first Bronze Star for this action. He would go on to receive two Bronze Stars, Combat Infantry Badge, Jump Wings, National Defense Medal and Good Conduct Medal.

Barnes came back home after his tour of duty as a Sergeant and spent his last seven months at Fort Devens, Massachusetts with the 10th Special Forces before discharge in 1971. He would find out that the person who replaced him was killed in action. He said he has no regrets going to Vietnam and was proud to serve but wonders what it was all for.

Back home, he used the GI Bill to go to college at the University of Baltimore, graduating from law school in June 1977 and passing the Bar the first time. Barnes began working part time in the States Attorney's Office as a law clerk in 1976. After one unsuccessful attempt running for States Attorney for Carroll County in 1989, he came back and won election in 1994 and has worked for the county ever since serving four terms to date and looking forward to a fifth.

He was part of the committee that helped build the Vietnam Memorial in Westminster. They started in 1989 and finished a year later. Today, Jerry Barnes still lives in Carroll County with his Wife, Florence (Flossy), who he married in April 2004. He has one stepdaughter who attends Stevenson University.

K. Douglas Beakes
Artillery Forward Observer
379th Infantry Regiment,
95th Infantry Division
U.S. Army
World War II

On the Lippe River in Germany, young Douglas Beakes captured five German officers in the basement of a building, disarming them. He directed artillery fire from the top floor of the house which was half full of American infantry. At nightfall, a German tank pulled up to and parked on one side of the house and began firing point blank on American troops scattered in the houses in the small town.

Beakes decided to call in artillery fire on the tank even though they were in close proximity to the American G.I.'s. The soldier called for a high explosive shell to land a short distance from the house. He then referenced the firing closer and closer to the house until he was finally able to order armor piercing shells to be fired that at last hit and disabled the enemy tank allowing him and the infantry to vacate the house. The house, which had been struck by numerous pieces of shrapnel, explosives, and tank fuel eventually burned to the ground.

Douglas Beakes was born November 17, 1923 in Washington, Pennsylvania but grew up in Baltimore, Maryland and Delta, Pennsylvania. Born of German and English heritage, he would become an all-around athlete and Eagle Scout. As an Eagle Scout he was selected to serve as a guide for foreign visitors at the New York World Fair and later as a guide to Civil War veterans of both sides at the "Last Reunion of the Blue and Gray" at Gettysburg in 1938.

Bill Hall

Beakes was a student at Western Maryland College in 1942 when he was drafted in the army. He, along with 18 other students from the college, boarded a special passenger car of the Western Maryland Railroad in downtown Westminster, Maryland, and headed for basic training at Fort Eustis, Virginia. After basic, he was selected for the Army Specialized Training Program (ASTP) and became a student in uniform at Westminster College in New Wilmington, Pennsylvania. When the program became disbanded Beakes was assigned to the 359th Field Artillery Battalion in Indiantown Gap, Pennsylvania.

After Divisional training, they sailed out of Boston for Liverpool, England eventually moving on to Weymouth, England and to a Landing Ship Tank (LST), winding up on Omaha Beach, Normandy, where he entered combat as part of Patton's Third Army. His unit would fight in France, Luxembourg, Belgium, Holland, and Germany.

When a forward observer became a casualty, Beakes volunteered to replace him, moving with attacking front line troops, crossing the Rhine River on a pontoon bridge and participating in battles in the Maginot Line, Siegfried Line and the Ruhr Pocket.

In November 1944 at the Battle of Metz, France, Patton needed a battle-proven unit, so he sent the 95th Infantry in. Metz was proclaimed to be the most fortified city in the world. The 95th proved him right. Even with the victory, the Division suffered 4, 000 casualties. General Keitel, Commander of all the German troops in the city, cited the men of the 95th as "the Iron Men of Metz." The name stuck and became the Division's insignia.

After Metz, they battled their way into Germany, becoming stalled on a high plateau along the Siegfried Line: the valley of the Moselle River in France to their rear and the valley of the Saar River in Germany in front.

In the freezing weather and deep snow, the troops were spread out. Out of nowhere, a German ME 109 fighter plane spotted the troops, banked, and returned spraying the troops with machine gun fire causing many casualties. They all knew it would return. As it returned, everyone grabbed weapons and began firing at the fighter. Beakes drew his souvenir luger pistol and fired a whole clip at the passing fighter.

A friend in his unit, Tony Vittiglio, heavy machine gunner, grabbed his weapon and began firing also. Vitttiglio had been a Black Shirt for Mussolini in Italy but fled to the U.S. in 1938 having no money, no English, and destitute. He was drafted into the U.S. Army and now fought for the U.S.

To the surprise of all the troops after Beakes and his unit began firing, the ME 109 lost speed, banked, drifted with wheels up and landed in packed snow, sliding until it stopped. The pilot threw open the canopy and jumped out firing at all the men around him. He was shot dead by the Americans. Beakes took a photo of the aircraft with a camera he had confiscated from a German house. His friend Tony had been acclaimed as the man who brought down the fighter from the sky but Beakes still thinks that some of his pistol shots hit the fighter as well. Today the plane is in a German's garage and he hopes to have it placed in a museum.

When the war ended, he was assigned to secure a hospital in Paderborn, Germany. It was used to provide retention facilities for German Africa Corps and SS officers who were suspected of war crimes. The prisoners used every opportunity to escape. Beakes fired the last three shots of the war and its aftermath. One German officer in full dress, descended the stairs, walked right past Beakes, and headed for the door. Beakes fired three shots over his head that splintered pieces of wood from the paneling. The officer stopped abruptly, turned, and walked right back upstairs. Beakes was eventually discharged December 22, 1945 at Camp Selby Mississippi. He received numerous battle stars, medals, and ribbons.

He returned to Western Maryland College and graduated in 1948, eventually earning his Doctorate. He worked for the Department of Defense for 46 years and in 1994 was presented the College Trustee Alumni Award from Western Maryland. He has published 17 books dealing with the Civil War, World War II, and the Cold War.

On December 4, 1009 in a ceremony at the French Embassy in Washington, D.C., Beakes was awarded the "Chevalier on the Legion of Honor." He married his wife, Maria, an Austrian, whom he met while skiing on the Alps. They have two children and five grandchildren. Douglas and Maria now live in Carroll Lutheran Village in Westminster, Maryland and became Carroll County residents in 2009.

Robert Boyer, Specialist 4
3rd Platoon, "A" Co. 2/12
25th Infantry Division
U.S. Army
Vietnam

It was a routine Search and Destroy (S & D) mission for young nineteen-year-old Bob Boyer. The Vietnamese weather exhumed its relentless heat and humidity on the American troops. In a wooded area during the daytime mission, they fired their weapons at suspected North Vietnamese troops. American Cobra helicopter gunships were called in to assist. The gunships began firing at the enemy, or so they thought. It was a day that would change the life of Spec. 4 Bob Boyer forever.

Born in Anne Arundel County, Maryland on June 1, 1949 to Norval and Marjorie Boyer, young Bob Boyer went to work after high school with Western Electric on an assembly line. The Vietnam War was getting hot and his chances of being drafted grew ever better. The day came in July 1969 and Boyer headed off to basic training at Fort Bragg, North Carolina.

While in boot camp, he became a hit with the recruits and the Drill Instructors (D.I.) doing his imitation of John Wayne. One D.I. even wanted Boyer to imitate him, which he did, and later gave the D.I. twenty-five push-ups.

Boyer then headed off to Fort McClellan, Arizona for eleven weeks of jungle and advanced infantry training. Afterward, he went to Oakland, California and on to Hawaii and ended up in Bien Hoa, Vietnam in the 25th Infantry Division. He stood tower guard duty for two weeks and moved on to Cu Chi which would be his home base.

His first ambush in Vietnam was horrible. His job was to hump the ammo for the machine gunner. He woke up during the night lying on his back with NVA tracers flying right over his face, making him freeze in the middle of the firefight. Boyer didn't even get to shoot but the machine gunner was firing away while Boyer lubricated the barrel with LSA so it would not get too hot. They left the area and when they came back later, they saw the dead Vietnamese soldiers, most of whom he said looked like children.

Boyer recalls how close the movie "Platoon" portrayed events in Vietnam, especially with the drug use. He remembers one night ambush when a cargo plane affectionately named "Snoopy," came in with machine guns blazing and firing all around them. To Boyer, it was one scary encounter. Their platoon did S and D missions mostly at night with a few during daylight. Most S and D missions lasted three to five days. On another mission, he was sent back with food poisoning and missed the action. During the night, the North Vietnamese Army (NVA) regulars walked right past his platoon.

Boyer was glad when he finally got a suntan because he didn't look like a new arrival. Once a month they would have a "stand down." It would be a picnic-type atmosphere with entertainment, steaks, and beer. The South Vietnamese would set up tables and sell everything from watch bands to cigarettes to massages.

On April 9, 1970 the platoon was conducting an S and D in a wooded area during daylight. Because of the firing they received, they called in Cobra helicopter gunships to assist. As the soldiers lay prone the gunships fired misdirected rockets at the American troops known as Friendly Fire.

The rocket struck about ten to fifteen feet from Boyer hurling several dreaded shrapnel pieces into his body and wounding several others. Both calves were damaged, the right more severe, and shrapnel hit his right arm. Part of his right thigh and calf were gone and his femur was broken. His foot was also damaged. Everyone began running away thinking another round was coming but when it didn't, they went back to assist the wounded.

The platoon called for helicopters to medevac out the injured, but the helos had to turn around due to the firefight. Finally, they came back. As Boyer lay there, he cried out to God to help him. While taking him to the helicopter, the stretcher broke and was quickly repaired while shooting was going on to protect the chopper. Boyer, the worst one wounded, was then taken to a MASH unit at Cu Chi. He never lost consciousness. He recalls the scissors cutting off his clothes and finally being knocked out.

He would find out later that after he was taken out by helicopter that same evening, "Zappers" came in at night and killed most of the men in his outfit.

He was sent to Japan for three months. He was put in a full leg and right arm traction. Boyer wound up getting a blood clot in his lung and was not expected to live but the clot finally dissolved. The young soldier spent months in full body casts and traction at Walter Reed Army Hospital. He made Specialist 4 while at the hospital. With a special brace, he was able to marry his high school sweetheart, Judy Boyer, on July 17, 1971.

Boyer received the Purple Heart, Good Conduct, National Defense Service medals and the Vietnam Service Medal and Republic of Vietnam Campaign Medal & Device. He was discharged in October 1971.

Bob and Judy moved to Carroll county in 1978 and have made their home here ever since. He recalls that day when he cried out to God to save him. He now says proudly that God has saved him "twice." He comes from a family with two brothers and two sisters and has a son and daughter.

Because of the severity of his wounds, Boyer has not been able to hold a job. He is a disabled veteran and now helps his wife of 38 years with her antique business. Although he has not been able to contribute physically, his mental attitude and courage have made him a standard-bearer for the veterans around the country who have suffered wounds in combat. If this man can take such a tragedy and turn it around for good, as he has, he has served his purpose and his country well.

Herbert Bricken
Lieutenant, U.S. Navy
Aircraft Carrier U.S.S. Franklin D. Roosevelt, CVB 42
WW II, Korean War

On board the aircraft carrier U.S.S. Franklin D. Roosevelt, CVB 42, Lieutenant J.G. Herbert Bricken was heading aft under the flight deck in a passageway. Suddenly, he was confronted with what every sailor has the most fear of on board a ship. Fire was in front of him. His instincts and training came to light quickly as he grabbed a fire extinguisher and began hitting the fire as best as he could, trying to hold it in check until help arrived. Finally, the damage control team arrived with hose lines and took over for him, eventually extinguishing the flames.

Bricken was born on June 7, 1926 to Barney and Sarah Bricken of Brooklyn, New York. They were immigrants from Russia in the 1890's. He had one brother. He recalls being smitten by seeing and touring the U.S.S. Saratoga, aircraft carrier, in New York as a twelve-year-old child. He also remembers getting sick while on board. After graduating from high school in June 1943 in Connecticut, he enlisted in the Navy reserves one day after his seventeenth birthday.

The young sailor was placed in the V-12 Navy College Training Program. The V-12 Program was created to complement commissioned officers in the navy during World War II. Most navy candidates went on a four-month course at a reserve midshipmen's school. He attended Holy Cross College in Worcester, Massachusetts in 1944 as a Seaman Apprentice. 14 out of the 120 men went into medical or dental training. He remained in college for twenty months.

Bricken then went to the Chelsea Naval Hospital in 1945. While there, he remembers marines being brought in from the Battle of Iwo Jima. He participated in taking off casts and bandages. He was horrified to find maggots, worms and infection once they were removed. The smell from the infection permeated his nostrils many times. But he is proud to say there were no amputees in his ward. He went on to NYU Dental School. After finishing the first half of his freshman year, he was mustered out and discharged in February 1946.

Bricken finished his freshman year at NYU, finishing dental school. He then transferred to the University of Maryland undergraduate dental school for three years. In his senior year he became a reserve ensign.

After graduating dental school, he joined the regular navy in June 1949 as a Lieutenant J.G.. He was sent to Parris Island, South Carolina for six months. While there, he was going to go home for Thanksgiving when his captain told him to stop in Washington, DC and see Rear Admiral Rault, Admiral of the Dental Corps. He told the Admiral that his captain said to do this. Admiral Rault asked him what he could do for him. Bricken said he would like to have sea duty. Rault asked if he would like to go on the carrier Franklin D. Roosevelt. Bricken just grinned! He went on board in 1950. He spent 28 months on board the Roosevelt.

The Roosevelt was the largest class carrier at the time. She was 986 feet long and could speed up to 33 knots. She also carried 3,000 plus crew. While on board he was promoted to Lieutenant. One incident he remembers clearly was when ammunition on board accidentally exploded. A crewman was injured by receiving shrapnel in the neck. Bricken was able to assist in the surgery but he also recalls that the crewman died from his injuries.

The Roosevelt toured the Mediterranean and the Atlantic. A moment he will never forget was while on a three-month cruise to Puerto Rico. While he was coming up to the main deck, the ship was coming out of the sun: the task force behind him. He viewed a fleet of about sixty ships, including five carriers, in the glare of the sun. He was so proud; it took his breath away.

An amusing story he tells is of a dental Commander who made a gold inlay filling for the Executive Officer, but he didn't want to tell the XO that it would not fit. He told the XO that a corpsman had thrown it out. The XO made all the corpsmen and two lieutenants, including Bricken, search the large container of garbage alongside the carrier for the gold inlay. He remembers how stinking the garbage was. All the while the filling was in the Commanders pocket! They all knew it.

Rumor also had it on board while in the eastern Mediterranean that they were carrying nuclear weapons as a deterrent to the Soviet threats to Turkey and Iran.

Bricken resigned his commission in 1953 to continue his studies and go to Periodontal School at NYU. In 1953 he opened his own practice in Silver Spring, Maryland and was certified in 1960 by the American Board of Periodontology. He had to go around and talk to many dentists because they did not know what he did. It was basically a new field. Bricken was #216 Board Certified in the whole country. He became head of the Periodontology Department of Sinai Hospital in Baltimore, Maryland from 1953-1958. He helped to reorganize the Washington DC area Periodontology Society in 1960 all the while lecturing at Georgetown and Howard Universities.

Bricken closed his practice in 2008 after fifty five years and had been one of the youngest Periodontologists in the country. He moved to Carroll County in 2008 with his bride of fifty-three years, Margaret Marilyn Bricken, whom he married in 1957. He enjoys his three sons and seven grandchildren while living at the Fairhaven Retirement Community in Sykesville, Maryland.

David Cammack
Lieutenant, Senior Grade
U.S.S. Valley Forge, CV 45
U.S Navy
World War II & Post World War II

The shakedown cruise for the U.S.S Valley Forge aircraft carrier CV-45 was over. Ensign David Cammack was experiencing a once-in-a-lifetime event. It was now part of a world cruise for the carrier and its crew. Now Task Force 38, Carrier Division 5 was called upon to do something that had not been done before. They were going to enter the Persian Gulf and become the first U.S. Navy task force in the gulf. Little did David Cammack know that shortly he would be feasting with royalty.

The future sailor was born to Howard and Sara Cammack on June 14, 1925. Howard was from West Virginia and Sara was from Philadelphia. His father was an insurance salesman and his mother a stage actress. Cammack had been in high school when he decided he wanted to go into the Naval Academy. He loved everything about it and it was his dream. That dream came true when he was appointed by his congressman and sworn in during the summer of 1944.

While at the Academy, he excelled in gymnastics for the High Bar. He also was the editor of the "Trident," the quarterly magazine for the Naval Academy. It was a literary and professional magazine.

Cammack was assigned to the Valley Forge in Philadelphia in 1948 after the shakedown cruise. They were to sail to San Diego and become the largest carrier to go through the Panama Canal. Their cruise would take Cammack to Hawaii, Australia, Philippines, Hong Kong,

Tsingtao, Singapore, India, the Persian Gulf, Gibraltar, Norway, England, New York, and return to San Diego.

One of his most exciting moments on board the Valley Forge was when he was Acting Officer of the Deck and in the Pilot House. While they were having flight operations they almost had a collision with a destroyer in the Task Force. The destroyer was going to a new position and didn't estimate the distance very well. Cammack recalls that the Admiral wasn't very happy. The ships missed by inches.

When they entered the Persian Gulf, they had orders to sail to Ras Tanura, Saudi Arabia. He believes it was because of the recent discovery of oil and the United States wanted to make a presence. When they arrived, the officers found out that they were invited to a special dinner. It was special because it was being hosted by the Crown Prince of Saudi Arabia, Abdul Aziz bin Abdur Rahman Al Saud, commonly referred to Ibn Saud, the future King of Saudi Arabia. The prince laid out a feast before their eyes. They felt very welcome.

As they steamed out of Gibraltar toward Norway, he felt the calling to explore becoming a minister. While stopped in England, he attended a beautiful service at Westminster Abbey. That convinced him that that was what he wanted to do.

After the cruise, he asked himself what made all the cultures he had seen tick. The trip around the world had convinced him to study for a career in the ministry. He also wanted to be able to be near his family and home. Cammack also remembers that when growing up he did not ever want to be a priest! He recalls that the Navy uniforms were much more attractive than priestly garments.

After the world cruise, the Valley Forge went back to operations on the west coast. Not knowing until later, Cammack had been picked while at the Naval Academy to be a teacher. He would go on to the U.S. Naval Preparatory School in Bainbridge, Maryland and be the head of the

English Department. He honed into his future career. While there, he was promoted to Lieutenant Senior Grade. He was discharged in the summer of 1956. He served 3 years at the Naval Academy and 2 years on board the Valley Forge. His remaining time had been in the Reserves. He would go on to receive the standard awards and ribbons for his service.

Cammack received his Master of Divinities at Virginia Episcopal Seminary in Alexandria and his MA at the University of Delaware in Human Relations. He was chaplain at the University of Virginia in Charlottesville, Virginia for 8 years and served 7 years as Directorship of Parish at Berryville, Virginia. He then moved to Baltimore as chaplain of local prep schools and served as half-time rector at a church in Woodlawn, Maryland. Eventually he became a full time rector at Trinity Episcopal Church at Waterloo before retiring to Dickeyville, Maryland. He moved to Carroll County, Maryland in 2002 and lives at the Fairhaven Retirement Community in Sykesville, Maryland.

Cammack married in 1949 while in the Navy. That marriage, after 4 children, dissolved. He would marry again in 1974 his wife Shirley. He has 3 daughters and 1 son from the previous marriage. The retired sailor now has 3 grandchildren. He has 1 sister whom he is proud to say married his Naval Academy roommate.

He loved his time in the Navy. It was a dream comes true from youth. He says today that he had pride and satisfaction in doing his job and was glad he did it.

William Dando
1st Lieutenant
Pilot C-47, Army Air Corps
5th Air Force, 317th Troop Carrier Group, 40th Squadron
World War II

March 6, 1945 1st Lieutenant William Dando was piloting his C-47 Skytrain over the island of Nasugbu in the Philippines. The crew of four had done this before. Their mission was to get supplies to American troops: mission accomplished. They were now heading to Clark Airfield in Manila. Flying at 1,000 feet to avoid anti-aircraft fire from the Japanese, they made their way across the island. Suddenly the ground fire became more intense than usual.

They were hit and started going down. Dando flew the mighty craft down onto the water filled with rice paddies. The airplane skipped, hopped, and finally settled into an embankment, finally coming to an abrupt stop. The crew, badly injured, never expected what would happen next.

William Dando was born on March 30, 1923 to Harris and Mary Dando, formerly of Frostburg, Maryland but who moved to Ohio along with his one sister. As a young man, he attended college in Akron, Ohio when after he started his sophomore year, like many patriotic young men at that time, enlisted in the Army Air Corps in October 1942.

Dando was processed at Fort Thomas, Kentucky and later went to Keesler Field in Biloxi, Mississippi where he took boot camp training. Because of his college, he was allowed to attend pilot training. First, he attended pilot training at San Antonio, Texas. Pre-flight was the first of the four stages. The second was Primary, than Basic followed by Advanced. The second and third sessions were held at Curtis

Field in Brady, Texas. Advanced moved him to Frederick, Oklahoma.

Finally, in the spring of 1944, Dando got his wish. He was sent overseas to Finschaven, New Guinea. Assignment for him was to the 5th Air Force, 317th Troop Carrier Group, 40th Squadron. Dando found himself assigned to fly C-47's, the workhorses, as the 1st Pilot. He would fly out of the Alandia Air Base.

The C-47 was vital in the Pacific campaign. They made it possible to airlift troops to counter the ever-moving Japanese army. They were famed for flying the "Hump" from India into China. In the Pacific theatre, they would use island landing strips to transport soldiers back to the U.S. mainland. They could transport up to 27 troops who would parachute into battle or they would supply ground forces with supplies such as food, water, and ammunition. And if needed, they would transport wounded troops to bases for medical treatment that could not be given in the field.

The aircraft did not have any weaponry. The only weapons on board were what the crew carried: a Thompson Sub-Machine Gun and a 45 pistol. Many times they worked with the OSS (Office of Strategic Services), forerunner of the CIA, by going into islands and supplying Pilipino guerillas with Thomson Sub-Machine Guns, ammunition and medical supplies. Most of the islands were occupied by Japanese troops but were of no real value in the campaign.

Dando recalls when they went into Mendora, they were the first American servicemen there since General MacArthur left. The natives were so happy that they brought out a bottle of Old Crow Whiskey they had been saving for such an event and drank it with the Americans.

In February 1945, Dando participated in the Battle of Corregidor. On February 16, troops loaded up in the C-47's on Mindoro and took off for Corregidor. At 0832 hours, the green light came on for the troops to begin their jump. Several others also dropped 75 mm pack howitzers.

March 6, 1945 was the day he will never forget. As they flew over Nasugbu, they were hit by enemy ground fire. As their plane glided across the rice paddy, the embankment stopped their forward progress, injuring the crew of four inside. One engine was torn off and the other was embedded in the mud. He could smell fuel leaking.

The Co-Pilot's cheek was fractured. The other two received minor injuries. Dando looked directly into the face of his knee and noticed that his foot was facing the other way. He knew they had to get out of the aircraft before they were captured but knew, with their injuries, it would be difficult. Dando also knew that they had crashed between Japanese and American lines. They were able to get out and tried to get to a stream on the other side of a hill. Suddenly natives came upon them and helped them to an American aid station. Morphine was the drug of choice for Dando who had suffered a crushed femur in his leg.

The only American doctor who was a dental surgeon wanted to cut off Dando's leg, but the injured soldier insisted it did not have to be. After 3 days in a schoolhouse/aid station, his leg was saved and he was evacuated to the 133rd General Hospital in Leyte, Philippines. While there, he recalls young Pilipino girls coming in every three hours to give him a shot of morphine in each arm.

Dando received the Purple Heart, Presidential Commendation Medal, Air Medal, and many others and received a physical discharge in September 1947, receiving treatments for two and a half more years for his war injuries. On Christmas 1943, he married his high school sweetheart, Janet, his wife of 62 years.

After the war, Dando worked for 30 years at the Libby Owen Ford Glass Company in Toledo, Ohio. He retired and worked in sales at various other positions and companies, and even started his own company at one point. He finished up his working life at Advance Glass Systems making bulletproof glass and heated glass for agencies such as the Secret Service and others.

After the death of his wife in January 2006, he decided to move to Carroll County, Maryland in October 2006 to be near his two children, three grandchildren and four great grandchildren. He now lives at the Fairhaven Retirement Community in Sykesville, Maryland.

Of his time in the service and World War II, he says he wouldn't trade it for anything.

Joseph Buford Davis
Staff Sergeant
Left Waist Gunner, B-24
Army Air Corps
7th Air Force, 11th Bombardment Group
98th Bomb Squadron (H)
World War II

Iwo Jima...D-Day +3. U.S. Marines were pinned down by Japanese artillery. Admiral Chester W. Nimitz asked for B-24 Bombers to be sent in but advised that they must be extremely accurate. Staff Sergeant Buford Davis was in the lead B-24 Liberator as the Left Waist Gunner. He would also take pictures with a large camera. They were given clearance for one pass over a large battleship to their targets. They knew never to fly over a battleship without clearance.

As they were ready to release their bombs, a large cloud bank enveloped the squadron. The pilot said they would make another run! The clouds cleared that time, and they dropped their bombs. They immediately thought of the battleship opening up on them, but it did not. On the way back to base, thinking of those Marines down on the island, Davis realized that they had to beat the Japanese. When they arrived back at Guam, Nimitz had sent a message commending them on their 90% accurate bombing run.

Davis was born July 7, 1922 in Birmingham, Alabama to Joseph B. and Ada Davis. He was an only child. He began attending the University of Alabama before eventually enlisting in the Army Air Corps in March 1942 at the Post Office in Tuscaloosa, Alabama. From there, Davis reported to boot camp in San Antonio, Texas. He was selected for Bombardier School where he took pre-flight training and from there went to Gunnery School. Because he did not score high on the gunnery targets, he "washed out" and wound up in Armament School. Davis had the chance to go back to Gunnery School, the same one he left and with the same instructor! That time, he scored 3rd highest in his class. He asked, and was granted, to be put back in Bombardier School and was eventually assigned as a gunner on the B-24 Liberators.

As part of the 11th Bombardment Group, they would see action in the invasion of the Gilbert Islands, the island-hopping campaign across the Pacific and participate in the Marshall Island campaign as well as the Marianas. They operated from Kwajalein Island eventually moving to Guam.

Davis recalls while on missions they caught a lot of flak, explosive debris cast off from shells exploding in the air around them. On one mission shrapnel struck the B-24 from below penetrating the craft between the pilot and co-pilot. They both received wounds to their right and left arms respectively.

They would fly low coming back from missions purposely to look for downed aircraft while subs searched on the surface. One downed crew, upon return to their base, raved about how great the food was on the sub that rescued them.

While on Guam, Davis got the scare of his life. As the men put up 5-man perimeter tents, they needed poles to put in the middle to support the tents. Most of the poles were bamboo from the jungles. Davis took a machete, headed into the jungle, and came upon an abandoned Japanese Sake garden. As he found a bamboo trunk he thought would work, he began hacking it when all of a sudden he felt like he had been shot in the neck by a sniper. He grabbed his neck in pain and, much to his surprise, found out it wasn't a bullet but a hornet that had stung him.

Davis also recalled that during the evenings, many times, the Japanese on the island would break into their mess halls for food. This also prompted the troops to put extra guards on their planes.

When moving their base from Kwajalein to Guam, they had one engine go out. Another caught fire and also failed. Their ball turret had also jammed and could not be opened. Two Navy Corsairs escorted them as they began throwing everything out they could to lighten the plane. Davis was pleased when they landed on Eniwetok Island and found out that they did not throw out the hidden beer they kept on the plane for their return. By taking it on the plane and flying at high altitudes, the air kept the beer cold. They had to wait for 2 engines to be delivered from Hickam Field in Hawaii. While there, a USO show was going on but they missed it. However, the star of the show, Betty Hutton, showed up to greet them because they could not make the show.

Another mission Davis recalls was when they were bombing a seaport. Their squadron caught heavy flak from the installations below. He even remembers when the B-24's flew "fighter escort" for Navy Reconnaissance Planes. All in all, he flew forty missions with the same crew. As a crew they voted to decline R & R.

He remembers being ticked off because everyone was celebrating VE day when a war was still going on in the Pacific and young men were still dying. With the war over, Davis was discharged in March 1946. He received several medals for his service including the Distinguished Flying Cross and the Air Medal.

The war veteran went back to college at the University of Alabama on the GI Bill and married Betty Davis. He majored in Political Science, changing to Law School but dropped out after one and a half years. He picked up his law degree at LaSalle University in 1970. Davis went to Atlanta clerking in a law office. He began working for Maryland Casualty Company working up to Assistant Division Manager. Davis wound up in Baltimore, Maryland. He retired in 1981. He volunteered at Sheppard Pratt Alcohol

Unit in Baltimore for almost five years and eleven years at Oak View Treatment Center in Ellicott City, Maryland.

Today, he lives at Fairhaven Retirement Community in Sykesville, Maryland where he moved to in 2006 with his wife, Betty. He has two sons.

J. Gary Davis
Captain, U.S. Army
A Co., 228th Assault Support Helicopter Co.
1st Cavalry Division
Vietnam

J. Gary Davis has a very real memorial from his combat in Vietnam. While flying a mission in his Chinook helicopter, his unit was hit four times by ground fire. Later they would find a metal plate behind a seat in the chopper with a hole from what they believed to be an AK-47 assault rifle piercing it. He cut the metal plate off the seat and took it home as a souvenir. Today, on his desk at work, he has a wooden model of a Chinook chopper on his desk. When looking at the bottom of it, one can see the sticker that reads "Made in Vietnam."

He was born September 11, 1946 to Warren and Charlotte Davis but because of his anger over September 11, 2001, he celebrates his birthday on September 12. After working in a machine shop as a teen, he decided to enlist in the Army in 1966 to avoid the draft. Davis' father served in World War II in the Army working ordnance in New Guinea.

The young soldier went to basic and advanced boot camp at Fort Gordon, Georgia and Fort Jackson, South Carolina respectively. From there it was to Fort Sill, Oklahoma for Officer Candidate School. After receiving his Second Lieutenant commission, he went on leave and got married to his wife of forty-two years, Dale.

Then it was off to Germany where he commanded B Battery, 2nd Battalion, 39th Artillery, 3rd Infantry. One of his proudest days was when he left Germany and went to flight school in 1969 and earned his "wings" and promoted to Captain. There, Davis received advanced training in the CH-

47 Chinook helicopter. The young officer was then off to Vietnam in January 1970. He was assigned to the 1st Cavalry Division and flew the chopper for the entire year accumulating 1,258 combat hours.

Davis recalls that there is nothing like flying a Chinook helicopter between trees and treetop levels at 140 knots. Then you throw in being shot at! The Chinook helicopter could hold up to 50-60 Vietnamese or normally carried an American complement of 34 and a crew of 5.

He was part of the first troops to go into Cambodia. In 1970, as they were crossing the border into Cambodia with some Republic of Vietnam airborne troops, they were listening to the Armed Forces Network radio and President Nixon's speech back home. Just as they crossed, he stated that our troops were now crossing the border to Cambodia. He thought that was really good timing.

The aircraft commander's craft was shot down once from the enemy firing into the chopper's rotors. They were able to get back to base and had the rotor replaced. After ten days in Vietnam, his chopper was to pick up a Connex Container, a large food storage unit, and take it to base. As they flew, the bulky container began a swinging motion. Piloting the chopper was very difficult as the container threw it off balance.

When they arrived at brigade headquarters, they were ready to put the load down and lost power at fifty feet. Davis could not release the load. As they came down, the back of the aircraft hit the load and put the nose of the Chinook into the mud. But the craft had enough movement to lift and land on its wheels. After settling, the rotors fell off. Later, the Safety Officer told him he could not understand how they survived with the engine gone.

Davis remembers a friend he met in San Francisco while waiting to go to Vietnam. He met Warrant Officer Leroy James Kowalski. They became very good friends. While on R and R, he found out that Kowalski's aircraft went down and he was killed. They would determine later that his flight controls disconnected in flight. He went to the Vietnam War Memorial in Washington, D.C. and found his friends name on the wall. For Davis, it was a very special moment.

For his service, he received 5 Bronze Stars (1 with a V), Army Aviator Badge, 48 Air Medals (1 with a V), Army Commendation Medal, Good Conduct Medal, National Service Medal, and Vietnamese Service Medals including the Cross of Gallantry.

One of the cruelest things that ever happened while he was in Vietnam was to his wife. She was told that there was a phone call, long distance, from him in Vietnam. When the phone rang, she picked it up but there was only a dial tone. It upset her because she did not know whether he had actually tried to call, if he was injured, or what had happened. She feared she had missed an important call but it wound up being a prank. They had discussed wives getting calls that their soldiers had been killed and he explained to her that if something like that did happen, there would be no phone call. A few days later her friend, not realizing what had happened, showed up at the door in his uniform. It scared her to death.

Returning home from Vietnam in January 1971 was a very special day for Davis. When he arrived, his wife, Dale, was all dressed up but had forgotten to take off her slippers! His two-year-old thought she had two Daddies when the cab driver helped him unload. His one-year-old kept pointing to his picture when asked where her Daddy was.

Back home, he went back to Fort Sill and taught gunnery to new officers coming into the Artillery for the first time. Then it was on to Fort Meade until his discharge in 1977. He and his wife moved to Carroll County in 1984. After the Army, he went back to college and received his degree.

The former soldier has four daughters and two granddaughters. He lives between Westminster and

Manchester, Maryland and is a CPA in private practice. He is also a partner in the Sturgill & Associates LLP in Westminster.

J. Gary Davis has two proud moments from his service in Vietnam. The first is that he had done all that he could do while he was there and the second was that all of his men came home.

Robert W. Delhamer
Captain, U.S. Army Air Corp
2nd Bomb Division, 355th Fighter Group, 357th Fighter Squadron
World War II

Young P-51 Mustang fighter pilot Robert Delhamer was leading his flight of 4 P-51's back to base after a combat mission escorting B-24 Bombers to destroy Hitler's hideout at Bechesfarten. As he switched from external to internal gas tanks, his fuel cock froze causing his engine to cut out. At 23,000 feet over Germany, he decided to try to land wheels up, saw a suitable spot in a field, and attempted a landing while his fellow pilots covered him. After the hard landing, he noticed vehicles moving quickly toward him. He quickly tore up his maps and exited the plane and headed for a grove of trees.

Delhamer was born on December 30, 1920 to John and Mary Delhamer of York, Pennsylvania. He graduated high school in 1938. After school he worked as a shoe clerk. He would listen to the news and believe that his country would be going to war very soon.

He decided to enlist in the Army Air Corps in August 1940, went through Boot Camp for six weeks and was classified for aircraft mechanic school. After satisfying the requirements for Air Mechanic 2nd class he was promoted to PFC. When the war broke out in December 1941, he was promoted straight to Sergeant.

Always wanting to fly, he passed the exam for flight training and went to Nashville, Tennessee for pilot training. In August 1942, he was promoted to the upper class as a Cadet Officer and then onto Arcadia, Florida for Primary flight training where he soloed 60 and flew hours including aerobatics, stalls, spins and cross-country flights.

Other schools and training would ensue such as Augusta, Georgia and Dothan, Alabama where he received basic and advanced flight training in an AT-6 aircraft. After graduation, he received his wings and was promoted to Second Lieutenant. Delhamer would also become a flight instructor since he qualified as an Expert Aerial Gunner as a student pilot. While at Napier, he promoted to First Lieutenant and assigned as Assistant Flight Leader.

After training he was sent overseas on the French Ocean Liner IL de France, arriving in Scotland. He recalls while on the IL de France sitting with Bing Crosby, the famous American singer. Then off to England and based at Steeple Morden after completing flight transition school in the P-51 aircraft.

Delhamer loved the P-51. It was the best single-engine fighter aircraft available during the war. It could speed up to 400 MPH and go to 40,000 feet. Its armament included six fifty caliber machine guns. He was assigned to the 357th fighter squadron of the 355th Fighter Group, part of the 8th Air Force. Their primary missions were to escort B-17's and B-24's on daylight raids on German targets. They intercepted various German aircraft, ME-109's, FW-190's, ME-110's and ME-120's in addition to the new ME-262 jet fighters. They would also seek out and destroy enemy targets of opportunity.

Their casualty rate was high particularly when attacking German bases. There were also casualties from loss of engine power on take-off or from mid-air collisions. Delhamer entered the squadron as one of five replacements. Three of those five were killed in action.

He survived two wheels-up landings: one in England and one in southern Germany. The first one was when he was in a steep bank in the landing pattern with his wheels-down when the engine cut out. He opened his canopy, retracted his wheels and landed wheels-up in a field, after barely missing some power lines. The propeller came off on contact with the ground and the left wing dug in causing the aircraft to cartwheel. The pilot struck his head on the gun sight causing unconsciousness.

The other was over German territory. As he got out of the craft, he was chased by troops on the ground before he realized they were Americans with Patton's Third Army. On another mission, he would receive the Distinguished Flying Cross for the destruction of 3 ME-262's on the ground on the same day.

Delhamer remembers leading a flight over Germany when a pilot cried out, "Jets…eleven O'clock high." They were the dreaded ME-262's. One got right on Delhamer's back and he could hear the bullets zinging past him. The American did a 360 degree turn and looked but could find no one. They were gone as fast as they came. During his flying time, Delhamer would fly 54 combat missions as a wingman, element leader, flight leader, and squadron leader. All in all he had 1,607 hours flying of which 267 were combat hours. He received numerous medals and ribbons for his service.

When the war ended, his fighter group would perform occupation duty until he returned home arriving in Boston. He served a total of 6 years and 5 months active duty but would spend time in the active and inactive reserves eventually retiring as a Lieutenant Colonel in 1980.

Delhamer would learn that his younger brother had been captured in Corregidor and was part of the Bataan Death March. He remained a prisoner for the duration of the war.

After the service, he would obtain a degree at night school at Johns Hopkins University and pass the test for certification as a CPA. He was employed for 37 years by the Johns Hopkins University, School of Medicine, serving as Director of Financial Affairs and Assistant Dean. He moved to Howard County, Maryland in 1973 and eventually to Carroll County in 1991. On Christmas Eve, 1942, he married Mary-Evelyn Delhamer, his wife of 67 years, and had 2 daughters and 1 son. He also has 3 grandchildren and 4 great grandchildren. While now living at the Fairhaven Retirement Community in Sykesville, Maryland, Delhamer still enjoys his hobbies of Astronomy, Geology, gardening, golf, bridge and furniture refinishing. Of his time in the service and his hair-raising incidents, he says that he is alive today because he was an instructor.

Richard Doxzen, Fireman 1st Class U.S.S. Stormes, DD 780 United States Navy Cuban Missile Crisis

Richard Doxzen was born to William H. and Dorothy M. Doxzen of Baltimore on November 23, 1941. Then in 1946 they moved to the Sykesville area on John Pickett Road, formerly Grimtown Road. Richard grew up a country boy working mostly as a farm hand. Shortly after graduating from Sykesville High School in 1959, and seventeen years old, he could not find a good job, so he enlisted in the U.S. Navy.

Richard went to Naval Station Great Lakes for boot camp. Following boot camp, he was assigned to the Atlantic Reserve Fleet and eventually to the Destroyer (Tin Can) U.S.S. Stormes which was in dry dock for a refit and Fleet Rehabilitation and Modernization (FRAM). A hangar deck was being added for a helicopter drone to be used in anti-submarine warfare. Doxzen was following his grandfather's footsteps. He had served on the U.S.S. Connacus, a minelayer, as a stoker. Richard became a boiler tender striker (someone learning boilers), assigned to the forward fire room just below the waterline. He would eventually attain the rank of Fireman 1st Class.

The Stormes patrol area was mainly the North and mid-South Atlantic Ocean. Its mission was to protect and patrol around the aircraft carriers launching Crusader Jets and locating foreign submarines by sight or contact. The carriers they shadowed were the U.S.S. Shangri La and the U.S.S. Randolph. Doxzen recalls catching a Russian sub refueling off a "fishing trawler." When they were spotted, they submerged.

Another special moment for the men of the Stormes was the recovery of Enos the monkey in a space capsule. Richard personally observed the recovery operation.

His duties were to keep the proper fire in the Babcock-Wilcox Boiler. His hours were usually four on and eight off. However, when battle stations existed, they were four on, four off and twenty-four-seven. Other times he would be in the upper handling room passing shells up an elevator to the five inch thirty eight mount.

With his tour of duty was now over and they were heading to Norfolk, Virginia from Guantanamo Bay, Cuba. He recalls the Stormes heading for Norfolk for service refit. He went to bed with the sun on the port side dipping low over the horizon. When he awoke in the AM the sun was just coming up on the "port" side. He immediately knew something was wrong. Then the (old man) Captain piped down that they were returning to Cuba to stop or block Russian ships from going to Cuba and that all hands being discharged were being extended for six more months in the service. Resigning to their dilemma, most of the crew said, "Let the Russkies come!"

While refueling off of the U.S.S. Randolph, the Randolph was hit by another ship which almost sliced into the Stormes. The Randolph wound up burning for one to two hours on the flight deck during the Cuban Blockade. The Stormes stopped at least one Russian freighter with missiles or supplies for the Castro regime. Because he was at Battle Station in front of the boiler, he did not get to see the freighters but heard the action over his "battle phones." They were off the shore of the Bay of Pigs and he recalls most of the crew thought they should go in and finish the job. But President Kennedy, after agreement with Khrushchev, declared the blockade over. Richard was discharged at Norfolk soon after in 1963.

The Tin Can Sailor recalls that after being on a bouncing destroyer at sea for three years and then suddenly being on solid ground that does not move was quite an extraordinary experience. Upon discharge, Richard met his wife, Norma Shipley, in 1964. They were married, raised two boys and now enjoy six grandchildren. He is retired from thirty years with law enforcement in Howard County, Maryland and currently drives a school bus which is an enjoyment in his retirement years. In his spare time, he is in charge of the Little Sykes Railroad, a train ride for young children from all over. It is located on Sandosky Road in Sykesville, Maryland.

William H. Doxzen
Sergeant, U.S. Army
17th Airborne Division
194th Glider Infantry Regiment
World War II

Sergeant William H. Doxzen will never forget his First Sergeant; nor will many others. First Sergeant Clinton Hedrick was twenty-six years old. As they entered the town of Lembeck, Germany, they came under heavy fire. Doxzen and others fought until the next day. Hedrick crossed the bridge into the castle and was met by a German self-propelled gun. Even though Hedrick knew he was fatally wounded, he kept firing long enough for his companions to seek refuge from the deadly fire. They all escaped, and the First Sergeant was awarded the Medal of Honor posthumously. Doxzen has never forgotten him.

William Doxzen was born on November 8, 1921 to Harry and Helen Doxzen of Baltimore, Maryland and was the oldest of seven children having three brothers and three sisters. His father was a veteran of World War I and his Dad's first captain was Harry S. Truman, future President of the United States. Young Doxzen attended School #85 in the city and from school he became a conductor on streetcar line #26 and #13 in Baltimore.

As World War II rolled on, he found himself drafted in the Army in 1944. It would be his home for two long years. After processing in Baltimore, he attended Boot Camp at Camp Blanding, Florida. For $55.00 more a month, he then volunteered for the airborne and took jump training at Fort Benning, Georgia. From there, he boarded a converted French ocean liner and arrived in Scotland. After receiving numerous shots, he and many others got sick but he does admit he never got a cold for about five years! He was immediately put in the Replacement Depot and assigned to the 17th Airborne Division. Then it was off to London.

The 17th and the 194th Glider Infantry Regiment of which Doxzen was assigned were supposed to participate in Operation Market Garden but it was instead passed over to the 101st and 82nd Airborne, citing inexperience and lack of supplies. After parachuting into France, they fought their way through Rems, Givet and charleville. On December 16, 1944 the Battle of the Bulge began.

During the Bulge, Doxzen recalls weather between 20 and 40 degrees below zero. It was the coldest winter in decades. He recalls the fighting as the worse. And at Bastogne, they put their fighting capabilities alongside the 101st Airborne and 87th Infantry divisions. The 17th had drawn as tough an assignment as they could get. During the first three days of intensive action, the 17th suffered about a thousand casualties a day.

Doxzen recalls one moment when he was sleeping in a farmhouse when a German 88 shell slammed into the house, totally covering him with debris. When he finally crawled out of the debris, his company had gone. He wound up catching up with them and asked them why they left him, and they told him they thought he was dead.

Another time he had hit a command car with a bazooka and tried to fire three more rounds. Finally, they fired but never blew up! Another memorable moment was when he and his friend, Sergeant Peterson, were walking along railroad tracks when a German tank fired a shell that did not blow up but bounced down the tracks between the rails. Once it stopped, Sergeant Peterson told him to walk slowly away from it!

He smiles when he tells how he got one over on a tank commander. They were sent as a platoon and three tanks to support a tank battalion. It had rained hard, and the grounds were extremely muddy. Doxzen got in the commander's tank with extremely muddy boots and went to sleep. He says he left the inside of the tank a "muddy mess." The Commander wanted to know who had been in his tank.

One memorable moment he recalls was when he was in a five-man patrol one evening and a patrol of at least forty German soldiers marched right past them.

Other battles they fought were at places named Wesel, where they glided in, Verdun, Dorsten, Haltern, Dalman, Appelhulsen, Lippstadt, Hamm, Essen, Duisburg, Mulhiem and Munster. Wesel was part of Operation Varsity. Munster was where he helped liberate a concentration camp.

He received a wound in the leg during his experience and years later had a piece of shrapnel work its way out of his left shoulder. He also recalls that after the war was over, there was a train wreck near Nancy, France. They assisted in pulling wounded soldiers out of the wreckage. While in Nancy he remembers seeing the lights of the city come on one evening for the first time in years. He recalls that it was an "awesome sight."

Doxzen says he lost many friends during the war, but the rule was to never get close to anyone because they may be killed the next day. His most favorite and memorable moment was hearing that the war was finally over and being able to take showers. Finally, in 1946, William H. Doxzen was discharged and allowed to come home after earning enough points. He received two Battle Stars, a Bronze Star, Purple Heart, Combat Infantry Badge and numerous others.

Returning home, he realized that he had some emotional issues to handle such as hearing slamming doors. Doxzen and his first wife moved to Woodbine, Maryland where he lived until 1969. He soon went back to operating streetcars in Baltimore for fifteen years and thirteen years in Washington, DC. Work then came from the B & O Railroad and eventually the Tennessee Valley Railroad as a conductor. In 1985, he moved to Rome, Georgia having married three times. At the ripe old age of 83, the old soldier decided to retire. Today, he visits often to enjoy his three children, five grandchildren and thirteen great grandchildren.

John (Jack) F. Feezer
Motor Machinist Mate 2nd Class
Landing Ship LSM 475
U.S. Navy
World War II

Motor Machinist Mate 2nd Class Jack Feezer never got seasick but now he was feeling very nauseous. His Landing Ship was packed with a crew of 60 men and 5 officers and in the after compartment were 70 troops and equipment. The waves buffeted the flat bottomed 302-foot ship from right to left and up and down like a cork in water in 30-35 foot seas. The great typhoon of October 4-9, 1945 had gotten in their way as they headed from Korea to the Philippines.

Their flotilla had been part of the invasion force, *Operation Olympic*, which would have begun the invasion of the Japanese homeland if they had not surrendered after the atomic bombs on Hiroshima and Nagasaki. He recalls one destroyer was lost. But in the scheme of things, they fared well. After the typhoon, the "Divine Wind" as the Japanese called it, sank 12 ships, grounded 222 and damaged 32. There were 47 personnel missing, 36 were killed while 100 were seriously injured.

Feezer was born February 2, 1920 in Sykesville, Maryland to John, Sr. and Frances Feezer and lived there with his two brothers and one sister. He left high school in his second year and began working full time for the U.S. Naval Experimental Station in Annapolis, Maryland. Due to the military-type work, he received a deferment but had a physical at the U.S. Naval Academy, nonetheless.

While there he worked in the test laboratory on a 2,000 H.P. submarine engine. He would test crankshafts to their max, log the information, and look for stress cracks and other anomalies that would cause the machine to fail: better there then out at sea or in battle.

On November 1, 1941, he married his sweetheart, Beulah Feezer. But fate dealt them a blow. He received his last deferment in March 1944 and was ready to be sworn into the Navy by the commander in charge of the test lab where he had worked. Three months later, in June 1944, all males under 28 had to go regardless. Feezer headed off to Boot Camp for six weeks in Bainbridge, Maryland.

Feezer took advantage of the schools. He spent six weeks at Basic Engineering School in Gulfport, Mississippi. From there he did six weeks in Advanced Diesel Engineering School at the same site. Then the sailor was off to Beloit, Wisconsin for six more weeks at Fairbanks Morse Schooling. He rounded things off at Little Creek, Virginia for three weeks at the amphibious base. His journey would then take him to war.

Young Feezer was assigned to the Landing Ship LSM 475 and loaded up at Gulfport, Mississippi. They traversed the Panama Canal and made their way to San Diego. On July 2, 1945 they arrived in Guam and ventured to Saipan arriving in August. The crew arrived in Okinawa September. 8, 1945 realizing that the atomic bomb had been dropped and the war was now over. On September 12, they left Okinawa for Korea arriving September 27. They were then met by one of nature's most ferocious events October 8, 1945.

Typhoon Louise had battered them and the flotilla unmercifully. They were 200 off Okinawa when a destroyer was lost. The 475 tossed and heaved in the pounding waves and winds over 100 miles per hour. The 35-foot seas crashed endlessly across the small ship. At one point the 475 almost capsized.

Finally they arrived in Manila only to be met by sampans loaded with native girls. Some of the crew stole sugar from the galley and gave it to the girls while others stole combat boots that were given to the crew and gave them also to the natives. When the Captain found out, he dressed the crew down very hard.

One time at General Quarters, Feezer was at his battle station in the Engine Room. He only latched one of the several "dogs" or latches to the hatch which led out of the room. An engineering officer came down and chewed him out because he did not latch all of them. Feezer immediately told him that if the ship got hit, he was out of there!

He thought very highly of the machinist shop on the 475. It had a worktable about six feet wide and twenty feet long. He enjoyed working out of there.

Feezer would wind up his tour in Jinsen, Fusan, and Samchok, Korea. He caught a mail plane to China and boarded the U.S.S. Roi, a converted carrier to San Diego, California. He then made his way home and discharged February 28, 1946. Today, he thinks about how bad it would have been had we had to have invaded Japan. The estimated American losses over that time were estimated to be a million men.

After the war, Feezer went back for two years to Baltimore Polytechnic Institute for his diploma and then continued night school for one year at Johns Hopkins University. He then opened up, with a partner, a store, *The Corner Supply*, in Baltimore County and worked part time for twenty years and full time back at the U.S. Naval Experimental Station until 1956. He sold out to his partner and started a contracting business, *Interstate Feezer Joint Venture*, a heavy excavation firm for twenty years.

He then helped develop property in Eldersburg, Maryland and then closed his business down. He also helped develop the "Junction" on route 32 near NSA. Feezer was on the Board of Directors from 1980-1990 for the Commercial Farmers Bank. Feezer is also a member of the U.S.S. LSM-LSMR Association #648. Today he still dabbles in stocks and remains active. He moved into Carroll County in 2002 with his precious wife of 68 years. Together they enjoy their children, 6 grandchildren and 13 great grandchildren.

Robert Fitz
U.S. Navy
Aviation Ordnance Man 3rd Class
Gunner, Avenger Torpedo Plane
U.S.S. Hornet
World War II

During the sea battle, a Kamikaze aircraft was trying to make its suicidal mark on the U.S.S. Hornet, CV-12 aircraft carrier. Eighteen-year-old Aviation Ordnance Man 3rd class Robert Fitz and a friend were under the flight deck of the carrier standing by a hatch. His friend said, "Let's take a peek." As they opened the hatch and peered out carefully, they noticed a Japanese Betty Bomber glaring down their throats as it dove toward the side of the Hornet. They quickly dived back in and slammed the hatch shut just as fire from the ship caught up with the bomber and it exploded before hitting the carrier. Luckily neither one was injured.

Robert Fitz of Waynesboro, Pennsylvania enlisted in the U.S. Navy in August of 1943. Born to Robert and Lottie Fitz on September 27, 1925, the young sailor spent his first eight weeks at the Great Lakes Naval Recruit Training Command Center in Illinois. From there he was sent to Shoemaker, California to Carrier Aircraft Service Unit 6 (CASHU 6). There they would service aircraft. Then he went to work in an office. While there he convinced a man to let him fly with him in a SNJ Trainer. They flew to Monterey Bay. On the way back, the man asked Fitz if he wanted to fly back. He was bitten.

The squadron received orders to ship out to Hawaii. Fitz asked the squad commander if he could go because he could fly. The commander told him when they arrived in Hawaii he would get two more planes and would then consider it. Then he was sent to gunnery school and finally assigned to the squadron. The same commander who granted his request would later be killed in action at Manila Bay.

Fitz was assigned to Air Group 11 on the U.S.S. Hornet in September 1944 and would remain there until February 1945. The Hornet (CV12) replaced the former Hornet (CV 8) that was sunk in the Battle of Santa Cruz in October 1942. He would become a gunner on the infamous Avenger Torpedo Planes. The Avengers had a crew of three: pilot, radioman, and gunner.

The Hornet saw a lot of action during his tour. Based out of the Admiralty Islands as well as the Uliithi Atoll, they raided many installations including Okinawa, Taiwan, Philippine Islands, China Sea and the Japanese Fleet. Aircraft from the Hornet protected troops invading Leyte in the Philippines.

During the Battle of Leyte Gulf, planes from the Hornet, along with Fitz, participated in damaging raids to the Japanese forces in Samar. His ship and aircraft destroyed an entire enemy convoy at Ormoc Bay. Aircraft also took photos of Okinawa in January 1945 that provided vital information for the coming invasion of the island.

At Clark Air Field in the Philippines, he shot down a Japanese Zero (Zeke) in the air. He strafed a lot of enemy aircraft on airfields during the campaigns.

His scariest moment came when the fleet was caught off Samar trying to take back Leyte. The Avengers carried one torpedo in their belly, secured by straps and a Bombay door. After the battle, the carriers were heading back for provisions and called back in early morning, catching the Japanese heading through the straits to the China Sea. They found two Japanese ships, one a cruiser, towing another wounded ship.

The Avengers headed for their targets. Taking heavy fire from the enemy as they glided in for the kill, Fitz and his crew sliced above the water and released their torpedo.

Nothing happened. The torpedo did not drop. The plane flew out of harm's way and they realized that they were in more danger than they had thought.

As they headed back for the Hornet, they tried to jettison the torpedo so they could land. Protocol usually would not let a plane land with a live torpedo in its belly. They would usually ditch the plane and rescue the crew. The Hornet was still landing planes.

As they could hear the humming of the live torpedo in its bay below them, the crew got the word to come in for a landing. As the Avenger slid across the deck, the torpedo went through the plane and slid across the deck as far as the superstructure of the carrier and never went off.

As Kamikaze's grew bolder, Fitz would feel the sting of war come close to him. A friend from the States on the carrier Lexington worked a gun station along a catwalk. A Kamikaze pilot struck the superstructure of the Lexington, wiping out three-gun emplacements along the catwalk where his friend was stationed. His friend threw himself down on the deck behind a steel barrier yet sustained severe burns on his back.

During its combat tour, Air Group 11 made a record destroying 377 Japanese planes and sinking more than one hundred thousand tons of Japanese shipping. For his service, Fitz received numerous medals and ribbons including the Air Medal and the Philippine Liberation Medal.

He was discharged in October 1946 and married Elaine Bentz, his wife of forty-two years until her recent death. He moved to Carroll County in 1996. Today he is married to Carol Walker of Randallstown, Maryland and lives in Manchester, Maryland with his two daughters, five grandchildren, and three great grandchildren.

He states that the U.S.S. Hornet is now a museum in Alameda, California yet one would think his home is a museum to his service on the Hornet also. Numerous ship and plane models and wartime memorabilia grace the interior of his home. He is a man who is proud of his service to his country.

Paul Garrett
Sergeant
17th Airborne Infantry
U.S. Army
World War II

The 17th Airborne Infantry was in Southern France about seven to eight thousand feet up in the mountains. It was the middle of October 1944. The local French told them they should leave because the snow was coming. On the 12-13 it snowed six feet. Sergeant Paul Garrett and his men, along with the 17th, retreated to the valley leaving two to three men behind to destroy equipment so that the Germans could not use them. The men left behind were wounded by bombs. Garrett and his men went back and saved their wounded comrades.

For years he never heard how the wounded had made out. In Huntsville, Alabama in 1990, Garrett attended a reunion and met one of the men he had saved. The man had lost a leg, one lung, and a kidney. It was one of the most emotional meetings of his life.

Garrett was born February 8, 1918 in Hampstead, Maryland to Arthur and Anna Garrett. Before the war, he worked at American Hammer making piston rings. But the twenty-three-year-old's life would change when he was drafted into the army on May 28, 1941.

Not knowing a thing about the military, he and a young man from Manchester, Maryland, were waiting for a train in New Cumberland, Pennsylvania and saw two MP's approaching. They discussed if they should salute them or not. They did. He attended boot camp at Fort Wheeler in Macon, Georgia. After boot camp, he heard of a new group forming but was not interested. It was for a Glider group. But history would change for him as someone else put his name on the volunteer list!

The army took 150 "volunteers" to Texas and trained them to be instructors in the 88th Airborne Infantry. They worked closely and trained with the 82nd and 101st Airborne Glider groups for one and a half years. They started out using planes but eventually went with the gliders. A glider could hold up to fifteen men but if a jeep or a gun were included, it narrowed space down to about six men. The airborne would go into an area behind enemy lines, control the area, and when the invading army went through, they would get a rest.

The 88th was dissolved into the 550th Airborne Infantry and in April 1944 they shipped out of Norfolk, Virginia to Oran, Algeria in Africa. They were there for ten days and then to Naples, Italy. On the way to Naples, a group became separated leaving them one company short. They then went south to Sicily and regrouped. One month later, they arrived in Rome and instructed groups in airborne operations.

In August 1944, they were loaded on gliders and flew across the Mediterranean Sea inside France ahead of the invading army. They took the first town in southern France, Le Muy. After about a week, they moved on to Niece, France and served as Military Police (MP's). They patrolled the city for two to three weeks.

After several months along the Italian border, they were moved to LeHavre in northern France and on December 16, 1944 loaded on boats and sent to England. It was the same day that the Battle of the Bulge began. Two days later they were back in France, then Belgium and Luxembourg relieving troops in Belgium that had taken the brunt of the Bulge. Some of the men associated with his company were captured in Bastogne. Most of the time, the army would keep troops in combat for no more than two to three weeks, pull them back, and let them rest.

Garrett and his men arrived in Munster, Germany, relieving a P.O.W. camp mostly filled with Russian soldiers. They were then set back to Paris where the 550th was folded into the newly formed 17th Airborne Infantry. By March 1945, they were loaded in gliders and towed into Germany. They landed two miles beyond the Rhine River, ahead of the British who were crossing the Rhine at the same time.

He and his men wound up in the Rohr Valley, an industrial area, pulling occupation duty. They conducted the dangerous duty of going house-to-house searching for weapons. One night when he and a partner were supposed to be relieved around 10 PM, a man from another company had had too much to drink and started shooting. Garrett found a bullet hole in his sweater between his side and his arm. It completely missed him.

Finally, the war was over. The army used a point system to determine if you could go home. To go, a soldier had to have earned 85 points. Garrett had 81 but finally got to 86 and was sent home in late September 1945 receiving his discharge September 23 with no scratches or wounds. He was married three weeks later to his present wife of sixty-four years, Geraldine Garrett.

After the war, Garrett went to work for Suburban Propane for thirty-five years. He is now retired, living in Hampstead and enjoys talking about his experiences. He has even contributed to a documentary about his life. He has one child, three grandchildren and four great grandchildren. His granddaughter's husband serves in the Navy in Virginia Beach as an F-18 pilot on a carrier. Of his call to duty, he said it was an experience, but he was just glad to be home.

Dennis Gore, Sr.
Gunners Mate Second Class
United States Navy, Naval Armed Guard
World War II

Living in Carroll County in the nineteen thirties was nothing like Dennis Gore, Sr. would find the rest of the world to be like. Born September 29, 1926 to Frederick and Alverta Gore, Dennis was born and raised in Carroll County. He worked a farm and drove a truck. But the peace and serenity of farm life would be shattered on December 7, 1941.

Soon after the attack on Pearl Harbor, Dennis' two older brothers enlisted in the military. Dennis would not allow them to go alone. At the ripe old age of seventeen, he joined the Navy in 1943. Dennis immediately volunteered for the newly formed Naval Armed Guard. This was a new concept where naval personnel were assigned to merchant ships and were trained as gun crews. Their mission was to protect the merchant ships from enemy ships and submarines. Prior to the Guard, numerous unarmed merchant ships were sunk carrying men and supplies across the Atlantic and Pacific by German and Japanese forces.

While in San Francisco undergoing training, Dennis would hear words he wished he had never heard. He was told of the death of his two older brothers. To add to his heartbreak, he could not even obtain leave to go home to be with his family because of the war.

The young sailor would serve in the Pacific making several crossings from the west coast to Hawaii and then on to Guadalcanal and the Russell and Marshall Islands Chain. While serving on the Liberty Ship Davy Crockett, Dennis was coming up onto deck from his Armed Guard quarters when an explosion suddenly ripped through the ship. The ship had struck a mine or had been torpedoed. No one knew for sure. The violent explosion blew Gore overboard along with others. As the ship went down, he and several others swam to a floating raft. Without any oars on board, the raft drifted toward the island of Saipan, then held by the Japanese. The crew was taken prisoner by the Japanese and held until the liberation of Saipan in July 1944. Dennis would suffer a shrapnel wound to his lower spine as a result of the explosion that went untreated and would haunt him for years to come.

Gore was assigned to the Seabees building a Saipan airfield while awaiting orders to be shipped out. The new field would accommodate the new B-29 heavy bombers. As they worked during the day to build the field, Japanese planes would bomb it at night. This continued until the arrival and intervention of the new night fighter P61 "Black Widow". Because naval gunners were allowed to volunteer to be gunners on bombers if necessary, Gore volunteered on a B-26 medium bomber. His plane returned from a mission with over fifty holes in it.

Nearing the end of his time on Saipan, he would watch B-29 bombers come and go. He could also watch the bombers on the nearby island of Tinian. One day he watched three B-29's take off from Tinian and sometime later felt and heard a "thump." Dennis finally received his orders to ship out and while on a liberty ship bound for Hawaii, the crew was informed of the bombing of Hiroshima. He then knew what the "thump" was.

With the war finally over, Dennis was discharged in 1946 and would be hospitalized several times due to shrapnel fragments left in his spine. Although he received ribbons for the Pacific Theater action, he received no medals; not even a purple heart. Like the Merchant Marines, neither the Navy nor the government recognized these men until years after the war's end. He relates that it was not a real problem adjusting to coming home. He did have a few bad dreams now and then but soon resumed his truck driving career.

Dennis Gore, Sr. now lives in Westminster, Maryland with his wife of fifty-four years and has four daughters and one son. In his words, he said that like so many others in that period of time, he did what he felt he had to do. His two older brothers went so he felt he should do the same. Dennis Gore, Sr. is another member of the Greatest Generation that lives on in our hearts and minds forever.

Sgt. Roland Gosnell
U.S. Army Air Corps
B-24 Liberator Top Turret Gunner/Flight Engineer
15th Air Force, 461st Bomb Group, 765th Squadron
World War II

On his next to last mission, Corporal Roland Gosnell, Top Turret Gunner/Flight Engineer on board a B-24J-CO Liberator bomber, with the ten-man crew were on their way to drop anti-personnel bombs on the Italian and German forces bottled up in the Po Valley region of Italy. The squadron of bombers turned back over the Adriatic Sea to fight the Germans.

Nearing the coast, flak began blackening the skies as enemy fire hurled into the heavens toward the crews. Without warning, the bomber directly in front of Gosnell's Liberator savagely exploded into a ball of fire, spiraling down toward the earth that patiently awaited them.

The tail gunner on Gosnell's aircraft called out that part of their own tail was missing. They looked out the waist gunner window and saw that the explosion had ripped the cowling off one of their engines, lodging it around their tail fins. Luckily there was no damage to the aircraft's rudder. Gosnell recalls that the real heroes of that mission were the Rangers who laid white bed sheets as markers for the bombardiers. After dropping their bombs, they finished their mission and returned to base.

Born in Carroll County, Maryland May 16, 1921 to Ernest and Birdie Gosnell, the future airman lived with his two sisters and brother. He started out working as a farm hand. From 1938 to 1941, he worked with a blacksmith and eventually at Locke Insulator Corp. in South Baltimore.

Drafted Sept 4, 1942, Gosnell did Boot Camp in Fort Meade, Maryland and then applied for the Army Air Corps. He was assigned as a cadet in Florida and passed the Cadet Board as a pilot. Pre-flight training came but was cancelled after two weeks. There were too many pilots! They were told the Air Corps needed navigators, radiomen, bombardiers, etc. Gosnell would become a Top Turret Gunner/Flight Engineer.

In February 1943, he went to Dallas Aviation School learning to be an aircraft mechanic and flight engineer. Assignment came to the 976th Bombardier Training Squadron maintaining aircraft for pilots and working the flight line. He was there for fourteen months. August to September 1944 he attended "Flexible Gunnery School" followed by Air Combat School from September to November 1944, rising to Corporal.

The crew picked up their aircraft in Lincoln, Nebraska, flew to New Hampshire, then on to Gander Bay, Newfoundland, ending up in the Azores. Next stop: Barie and Gioria, Italy. While on their way across the U.S. to New Hampshire, they flew over Indiana where some of their families were. They all put notes and letters in the bottles for their families. The pilot broke all the rules. He changed course without telling Controllers, dropped to about three hundred feet, opened the bomb bay doors and dropped the bottles out. Gosnell's wife, Leona, got hers three weeks later.

Assignment came to the 461st Bomb Group, 765th Squadron. In Barie, Italy, the soldiers and airmen were warned that the town was off limits and they were restricted to base. Three of his crew decided to go anyway. The Military Police (MP's) arrested them for impersonating an officer (one had a Lieutenant's jacket on), carrying illegal money, and carrying a concealed weapon in a gun free zone. Their officers were told they would keep the crew together because they were going on a mission the next day. They were then ordered to sign papers for the three men to be court-martialed in the States when they got back. While over the ocean the next day, they opened the bomb bay doors and threw the papers out into the ocean. They never did hear any more regarding the incident!

One thing that Gosnell remembers is the unwritten code of conduct between airmen on both sides in Europe. If an aircraft was damaged and lowered its wheels, no one would attack you. The same was with fighter aircraft. But he warned that if a man remained in a turret, he was fair game because they thought you would shoot at them.

The airman recalls the worse campaign in Army Air Corps history. Between April and August 1944, they lost 325 planes (ten-man crews) over Ploesti, Austria while bombing oil refinery plants. He also praises the Tuskegee Airmen (Red Tails) who gave fighter support for their bombing missions.

His most memorable moment was crossing the one hundred fifty miles of the Adriatic Sea with a full bomb load, low on fuel, and structural damage knowing that if they had to bail out, they would do so on the shores of Yugoslavia. None of his crew was injured, though losing an engine and physical damage over the course of thirteen missions on a B-24 and seven on a B-29.

Gosnell recalls saving two dogs by hiding them in duffle bags on board a ship while taking them home. The captain of the ship wanted to throw them and the other "30" dogs overboard due to regulations but was eventually talked out of it. The two dog's names were Scotch and Soda.

He eventually was discharged October 26, 1945 but remained in the Air Force Reserves for nine years. He received the ETO medal with 4 battle stars, Air Medal, Unit Citation, Combat Action Ribbon, WW II Victory Medal and Good Conduct Medals from both branches.

Many jobs followed after taking advantage of the GI Bill and getting his pilot's license and helping the Civil Air Patrol. He spent many years at the American Can Company and started his own Rogoco Engineering Company. Today, he lives in Eldersburg, Carroll County with his two boys and daughter, three grandchildren and six great grandchildren. He was married to Leona Gosnell for sixty-five years until she passed away two years earlier. To pass his time he builds RC airplanes and boats and is a certified machinist and gunsmith.

Walter Gray
Private, U.S. Army
4th Armored Infantry Regiment
2nd Armored Division "Hell on Wheels,"
9th Army,
World War II

The last thing Walter Gray wanted to be was be a Prisoner of War. When they got to the Elbe River they were told to stop. 150 men crossed and went to within fifty miles west of Berlin. There they stayed for two days and three nights with no action. Relaxed, the men fell asleep. When Gray awoke, his men were holding their hands up in the air with three German tanks pointing their guns at them. With all he had in him, he got up and ran, expecting to be shot in the back. He ran and got into a bunker over a hill, met two other G.I.'s, crossed the river and entered the town of Magdeburg on the Elbe River.

The Penn Hills, Pennsylvania native was born to William and Annie Gray on November 21, 1925. He was drafted August 1, 1944 and entered boot camp at Camp Joseph T. Robinson in Little Rock, Arkansas for fourteen weeks, finishing in December and spent Christmas in Baltimore, Maryland. He was trained in armored infantry. His pay was S57.00 a month with an extra S10 a month for overseas duty, another S10.00 for combat for a grant total of S71.00 a month.

On January 1, 1945, he was shipped overseas, arriving in LeHavre, France. From there he went on a train to a replacement depot in Belgium and was assigned to the 2nd Armored Division, 41st Armored Infantry. He would see the last half of the Battle of the bulge. Gray saw many of his friends and comrades killed in action. To this day he calls it his scars. The soldier relives those events once in a while but then they fade away, recalling that at first it was terrible.

During the last half of the Bulge, they replaced the dead and wounded. One soldier went overseas with him and into his company. They were in the rifle squad guarding tanks. The Germans shot through the edge of buildings and took friends head off. One Italian American soldier ran across Italian soldiers heading back to Italy. They encouraged him to go with them, but he refused saying he was an American. He was killed the next day.

They finally crossed the Siegfried Line into Germany and came to the town of Achen. He was mesmerized by the destruction but was amazed that in the middle of the devastation sat, untouched, a F.W. Woolworth's Department Store.

The young private was assigned to a 30-caliber gun on the back of a Half Track vehicle. Other times he would operate a 60 mm Mortar. Because of the sixty-pound weight of the mortar and ammo they had to carry, they would often swap with each other. The easiest part was carrying the binoculars.

Many times, they would have to level towns along the way due to resistance. Then they would stay in what was left of the homes. Gray and others learned to speak German very well in their short time there. A squad leader from West Virginia approached a farm saying he was going to get some real eggs. A German man saw him and the soldier spoke in broken German what he was going to do. The German wound up being from Chicago and told him to speak English if he wanted eggs!

They released a lot of Polish and Russian prisoners along the way. They once tried to give them rations but were told to stop because the very thin and starving prisoners would get very sick and possibly die.

Even though they crossed a pontoon bridge with very little on each side of the vehicles other than water, the young lad declared it was hell crossing the Elbe River. The worst soldiers were the SS troops. One town nicknamed the SS Valley was terrible. Mountains were on both sides with trees on one side and a Pasteur on the other. The German troops poured machine gun fire on them until a G.I. sniper, one by one, got them.

One soldier, a Sergeant Wilson talked about getting home and marrying his girlfriend. He got a Dear John letter and was devastated. While under fire, they were in trenches in a roadway. Wilson kept popping his head up in the same place and a German put a bullet right through it.

With the war over, they became occupation forces. One thing Gray hated was guard duty. To get out of it he drove a ¾ ton truck and was promoted to Corporal. Truck Drivers got out of guard duty. He was able to drive a lot through Germany. Volunteers were needed to take displaced Polish troops back to Poland. He would leave Nuremburg and go to Prague, Czechoslovakia and on to Poland with a return to Germany. Later, on a trip to Paris while driving a captain, the jeep they were in hit a sand pile and went into a ditch, driving into a telephone pole. Gray was thrown through the roof cover but both survived.

He recalled how when he got home, he didn't know how he was going to let his mother know he smoked cigarettes; after all, they were fifty cents a carton! She calmly told him to go ahead. He had earned them. Along with his parents, his two sisters and brother welcomed him home. He was discharged with several medals and ribbons on July 1, 1946 without a physical scar.

He moved to Carroll County in 2001 to be with his family. Today, after working U.S. Steel, the retired soldier is now Chaplain of the American Legion Post 223 in Sykesville, Maryland. He lives in Eldersburg, Maryland with his wife Joanne, three daughters, six grandchildren and three great grandchildren.

Robert Hale
Technical Sergeant
972nd Signal Corps
U.S. Army
World War II

Robert Hale was born November 11, 1921 to Robert and Eleanor Hale of Baltimore. He was the seventh Robert Hale in his family tree. Robert had one brother. When he was seventeen, he joined a month-long summer military training camp at Fort Meade, Maryland called Civilian Military Training Corps. They were taught military courtesy, when to wear different military uniforms, how to march, care for weapons and how to shoot them along with use of military radios and telephones.

In the fall of 1941, he entered Johns Hopkins University and took ROTC as one of his courses. Then the war began on December 7, 1941. In the summer of '42, he received his draft notice. His ROTC Captain suggested he enlist instead of being drafted. He reported on December 7, 1942 to Fort Meade. In January 1943, he went to Basic Training at Camp Crowder, Missouri, which was only one month in the Signal Corps.

Hale was sent to Central Office Wire Chief School and promoted to Corporal. This taught him how to run and operate a small telephone system such as in a small town or base. He then went to Fort McDowell on Angel Island in San Francisco for deployment overseas. He was sent then to Ft. Shafter Honolulu, Hawaii where he was assigned to the 972nd Signal Battalion. The 972nd Signal was attached to Headquarters Pacific Ocean Area and had personnel on duty at different bases in the Pacific. When the Colonel saw Hale had had a surveying course at Hopkins, he was told they desperately needed surveyors.

In June 1943 Hale was sent to the big island of Hawaii on detached service for seven months where he was furnished a survey crew, rations, and quarters by the 245th Signal Construction Company. Returning to Ft. Shafter In February 1944 he was promoted to Sergeant and was placed in charge of a survey section consisting of four civilian survey crews working all over the island. November 1944 promotion came again as Staff Sergeant. He joined an engineering group consisting of four officers, four civilians and nine enlisted men that was heading to Okinawa to design a permanent communications system for the island.

They stopped at a couple of islands and then Guam. When they approached Guam, their plane circled, and the pilot notified them that their instrument panel indicated that the landing gear was not down and locked. They were told they were going in for a crash landing. As they approached, they saw fire trucks and ambulances driving alongside the plane. When they landed, the pilot realized the gears were down and began to slow the craft down causing the left inboard engine to catch fire. The onboard extinguisher put it out. That was when they found out that the cargo compartment contained proximity fuses for bombs. Everyone was okay.

When they arrived on Okinawa, there was still fighting going on. As they approached, they could see the flashes from the battleships firing shells onto the island. The island was ninety percent secure, but a lot of fighting was still going on at the southern end. Hale remembers a bomber coming over at ten, two and five o'clock every night just to let them know they were still there. He was amazed at the amount of stars in the night skies over the island during a blackout.

Everyone was aware that there were many enemy deserters on the island hiding from the Americans. Every day he would get G2 reports. One day he was walking down a narrow path and passed a demolished building with a thatch roof. He had his map and compass in hand when he heard a sound behind the building. Instantly he dropped the equipment and drew his pistol faster than the fastest gunslinger. He tried to decide if he should back up, go forward or investigate. As the hair stood up on the back of his neck waiting to see the enemy, a goat stuck his head out of the debris in the area. Relieved, he burst out laughing with no one to hear him.

Promotion to Tech Sergeant came next as the famous typhoon hit Okinawa in 1945. Winds in excess of 150 mph pounded the island. After about twelve hours, the eye of the storm came over them with a beautiful clear sky for a few minutes. Then the remainder of the storm hit. The men were so tired from the first twelve hours that they slept through the rest of it.

After the storm the island was a wreck. Hale's job was to patrol a certain section of the island to check the condition of the telephone pole lines. Only the Corps of Engineers, the Signal Corps, Quartermaster and Medical Vehicles were allowed to get gas as those units needed to rebuild the island. He remembered getting five-gallon cans of gas that had been buried for just such a problem.

On December 14, 1945 he received orders sending him to Fort Meade to be discharged. After thirteen days, he arrived in Seattle aboard the More McDove. Waiting for him was a woman in a Red Cross uniform. It was his cousin. He stayed with her one night before going to Fort Lawton to wait for transportation to Fort Meade. He received his discharge on January 16, 1946. He had not been home for three years, one month and nine days.

Two years later, he was given a commission in the Reserves as a Lieutenant where he spent nine years in a reserve Signal Corps unit.

Bill Hall

Over the years he worked as an Electrical Engineer for General Electric, Westinghouse and Century Engineering before retiring. He moved to Carroll County in 1998 with his wife Joan. They have been married fifty-six years. He has a son and daughter and three grandchildren and enjoys Ham Radio as a hobby.

Robert Hall
Corporal, U.S. Army
45th Infantry Division (Thunderbird),
180th Regiment
Corps of Engineers
Headquarters Company
Korean War

June 1, 1953 Corporal Robert Hall was anticipating going home in a little over two weeks. His unit was advancing up a rugged hill, using ropes, to reinforce the main line of resistance. The left flank had fallen behind them making them susceptible to be trapped. As they got to the top, a friendly fire shell came in and landed below him, blowing up. All Hall remembered was an orange flash. His body was thrown into a door of a bunker.

As he gathered himself together, he began to feel his body to make sure he was okay. He went outside and looked back at the door. To his amazement, it was like an angel arc that surrounded where his body had been. Shrapnel had torn holes throughout the door, missing him completely. To Hall, it was a miracle. His sergeant was not so lucky. Staying up to show a new sergeant the listening posts, he was killed, losing half his head and half a leg. Hall carried him off the field.

Robert Hall was born in New York City October 16, 1930 to Harold and charlotte Hall of New York. He had one brother who had joined the Marines and later the Army. During the day he was a hotel maintenance mechanic and in the evenings he attended New York City College.

In June 1951, Hall was drafted into the Army and headed off to Fort Dix, New Jersey for sixteen weeks of Basic Training followed by eight weeks of Infantry Leadership School. Then he was off to Fort Belvoir, Virginia for Corps of Engineer training for eighteen weeks. There he learned mines, explosives and demolitions.

Assignment finally came and he was sent off to Korea. He departed San Francisco and flew on a DC 4 to Hawaii and then Wake Island, bouncing through Tokyo and landing in Pusan on a C140. He took a train to Seoul and in June, 1952, he was assigned to the 45th Infantry Division and placed on the line. As he arrived, he could see a U.S. Navy ship firing shells inland and thought of all those sailors and if they died, they would die clean unlike the grunts in the foxholes. His tour would be for one year.

Hall recalls the fighting in Korea was similar to WW I. Fighting was done mostly in trenches. He also remembers that the worse part of the living conditions was not being able to take a shower. One time, he didn't take one for three months. They literally lived in a hole in the ground. The conditions were so abhorrent he remembers a Chinese soldier that he had killed being left out on the field of battle for a month.

In his first half of his tour, he would fight at Heartbreak Ridge and Pork Chop Hill. For Pork Chop, his unit received a Presidential Unit Citation for rescuing the beleaguered American G.I.'s there. His unit set a combat record for 200 hundred days on the line. He is proud to say the 45th never lost an inch of ground.

While placing white tape across a mine field to designate clear areas, a Communist Chinese sniper began firing at him. Hall hit the ground every 50 feet or so and safely got to the trench line. The 2nd Lieutenant told him to get back out there because the job was not finished. Hall yelled to the officer to look out. A shell had come within inches of the 2nd Lieutenant's head. The officer finally agreed with him. They looked across the field and saw the white tape line and realized that they could not go out during the day, never used white tape again, and switched to black communication wire with white tape every few feet.

While on the front line, a priest was performing mass for 12 attendees when a White Phosphorus (American) round hit and landed right in front of them. Hall walked over to the shell, picked it up and tossed it over the hill and was surprised to hear it explode! He became very religious during the war. While on a five-day R & R in Tokyo, he was confirmed by the Archbishop of Tokyo.

Hall recalls one time while they were out on patrol, South Korean Troops began firing at them not realizing they were Americans. Luckily, they were bad shots.

Robert Hall was discharged July 31, 1953 but remained in the Reserves for six additional years. For his service he received the coveted Combat Infantry Badge, UN Defense Medal, Korean Campaign Ribbon with 2 Battle Stars, Good Conduct Medal and the Republic of Korea Presidential Citation. He would achieve the rank of Sergeant First Class.

Hall remembers the "Forgotten War." Most of his friends didn't even know he had been gone. When he came home, they asked him where he had been. He went back to school at Pratt Institute in New York earning his Engineering degree. He says his wife earned her "PHT" (Putting Husband Through) degree. Hall worked for IBM for 11 years and went on to be Assistant Vice President of Irving Trust on Wall Street for 17 years, advancing eventually to Executive VP and CIO. He moved on to Prudential Asset Management Company for 10 years as Senior VP and CIO.

He volunteered with U.S. Aid assisting retired executives around the world. Hall eventually went to Ryad, Saudi Arabia to help them in setting up the Saudi Arabian Central Bank. For 21 years he lived in Chatham, New Jersey volunteering as a volunteer EMT for an ambulance squad.

In 2006, Hall moved to Carroll County, Maryland with his wife Alice of 57 years to be closer to their two grandchildren. He also has three daughters, one of whom was killed in an automobile accident at age 17.

Lt. Colonel Ronald Hollingsworth
25th Infantry Division
United States Army
Vietnam War

From Francis Scott Key High School to Morgan State College graduation with a BS in Math and ROTC training, Ron Hollingsworth probably never thought he would see combat in any war. But Vietnam was brewing and escalating in intensity, patiently waiting for him.

After receiving three draft notices, the twenty-two-year-old would leave the quiet life in Union Bridge, Maryland and his parents, Charles and Eloise, to become a 2nd LT in the Army in June 1964.

The army wasted no time with this young man. He was assigned to the 25th Infantry Division. Upon arrival to Vietnam after a seventeen-day crossing, they stormed the beach to find the army band playing music. He was sent directly to Operations who gave him his orders to proceed to Cu Chi, northwest of Saigon. A major assigned him to a mission with a South Vietnamese battalion as Field Artillery Forward Observer. The mission lasted three weeks and when over, they had lost one half of the battalion.

After two days, he was reassigned to the 1st of 27th Infantry Battalion. There he said the world erupted and involved two hot fire fights. Half of them were wiped out. After ten days, they went back to HQ. Then it was an armored cavalry squadron, B Troop, which went to rescue a helicopter that had been shot down. Within four miles of the chopper, they were ambushed and again within two miles. A few tanks were knocked out. VC troops were climbing onto the Armored Personnel Carriers to get at them. Many men were lost.

He continued to cover convoys between Cu Chi and Saigon. After being promoted to 1st Lt, he was training mortar crews. Ron was sent to the 3rd of 13th Field Artillery which sported 8" guns and was the executive officer in charge of all gun crews.

Probably the most memorable battle was that of Nui Ba Den (Black Mountain) near the Cambodian border. The 1st of 27th Infantry Battalion was being hammered by VC on the mountain and called in for help. Hollingsworth's crews fired 8" shells for ten days continuously in support of the troops, eventually being supported by three American Divisions and one brigade. He recalls looking around after the battle and seeing a junk yard of empty shell casings lying around them. His crew had saved the 1st of 27th and for their actions, several, including Ron, were given the Bronze Star. Two months later, he came home.

When he arrived in the U.S., while waiting at Chicago airport, a man started yelling at him calling him a "baby killer." People were telling him things that were happening, and he let them know it was not so. But the hardest part for him coming home was seeing his wife and daughter at the airport waiting for him. His young daughter would not go to him. It took a few weeks before she remembered who he was.

Ron became a gunnery instructor at Ft. Sill, OK and was promoted to Captain. He then was sent to the Officers Advanced Course during which he was taught gunnery by one of the majors that he had trained. Then it was to Vietnam for a second tour. He went back to the same area, the 23rd Field Artillery Group, and assigned to the 1st of 12th Field Artillery Battalion as Fire Direction Officer, eventually taking command of B Battery. Here he was wounded when his unit was attacked. Later he became a Fire Base Commander. This was command of his own Battery and several of the armed forces stationed there.

The soldier went to Fort Meade, Maryland for four years at the Inspector General Office, helping to resolve racial issues in the Army. Ron attended the Command and General Staff School in Carlisle, PA and Ft Leavenworth, KS. Promotion followed as major and commanding officer at Ft. Miles, Delaware. He took two tours to Germany having specialized in computers.

Command of a battalion in the 3rd Army Division came soon after and a promotion to Lt. Colonel. This led him to command of the 1 of 40th Field Artillery Battalion in Hannau, Germany. The officer decided to retire in June, 1984, having led a full life while serving his country. For his service, he received the Purple Heart, two Bronze Stars, Air Medals, Vietnam Campaign Medal, and the Vietnam Service Ribbon.

Hollingsworth would not retire fully. He became a software engineer for Martin Marietta, eventually becoming a department manager. There he worked for eleven years and opened his own business as a software consultant, finally retiring to take care of his bride. He still dabbles in real estate today.

The retired soldier now lives in Westminster, Maryland with his wife Shirley, of forty-seven years. They were married in 1962. He has two daughters, Lori Ann and Ronia and two grandchildren.

John Jennings, Jr.
2nd Class Petty Officer
Quartermaster
U.S.S. Greenwood, DE 679
World War II

On May 18, 1926, John Jennings, Sr. and his wife Nannie Lorraine gave birth to a three-pound premature baby boy. He would be named John, Jr. after his father. John, Jr. would grow up to hear of places like Pearl Harbor, Guadalcanal, Midway, Coral Sea, and Wake Island. While in Sparrows Point High School, he dreamed of serving his country in the Great War.

In May 1943 at the age of seventeen, he enlisted in the U.S. Navy and, after boot camp in Bainbridge, Maryland, was assigned to the U.S.S. Greenwood, a Destroyer Escort vessel, as part of the original crew. The ship would be his home for the next four years. As a crewmember, John wanted to be a Pharmacists Mate, but an option opened up and he began "striking" for Quartermaster. Striking is a navy term for someone who is learning a job. He would wind up excelling as Quartermaster eventually achieving the rank of Petty Officer Second Class.

On his first time out at sea, John's ship hit a hurricane. As they tossed, dipped, and bobbed, the crew below decks kept hearing a rolling and banging sound. Later they would find out that it was a depth charge that had broken loose and was rolling around on the deck above them. Lucky for them, it was not armed.

The Greenwood headed out for the Pacific in November 1943. She reached Samoa in December. In January 1944 they arrived at Guadalcanal. There, for the first year, they served as escort for various ships through the dangerous waters of the Solomon and New Hebrides Islands. Other duties included anti-sub sweeps and firing depth charges.

In January 1945, the Greenwood and John joined the 7th Fleet at Leyte Gulf. After escorting Merchant Men, Troop Transports, and LST's to the Philippines for the initial landings at Lingayan Gulf, they provided screening and anti-aircraft fire in support of the fleet. From there, John and his crew sailed for Ulithi, an atoll in the Caroline Islands, to pick up a convoy of supply and troop ships bound for the invasion of Iwo Jima. Not long after their arrival, the Greenwood learned that the ship they had relieved had been sunk.

While at Iwo Jima, John could watch Marines crawling up the island and throwing grenades into caves. They provided fire support and escort to other ships. Some of her other duties were anti-sub patrols, where she was involved in several depth charge attacks, beating back Japanese air strikes, and providing direct fire on the island in support of the troops. The Greenwood was also used as an illumination ship. They would fire star shells all during the night to help Marines spot enemy night attacks and infiltrators. After a month on scene, they left for the Eniwetok Island providing escort for transports loaded with wounded Marines. The Greenwood, also, carried several wounded soldiers. One of his most haunting memories of the war was, as a young man, holding a young, wounded Marine in his arms and watching him die. To this day, it still hurts him to think about it.

In May 1945, John and the crew of the Greenwood began operating with aircraft carriers in preparation for Operation Olympic, the invasion of the Japanese homeland, but operations were halted after the bombings of Nagasaki and Hiroshima in August of 1945.

The war was finally over. One of the most memorable moments for John was coming home. As quartermaster, he was helmsman of the Greenwood. As they sailed under the Golden Gate Bridge in San Francisco, he steered the Destroyer Escort through the Bay. He looked to port and saw a huge sign on a hill reading "Welcome Home U.S.S. Greenwood DE 679." To him, it was one of his proudest moments.

His remaining time in the Navy was on the Greenwood doing escort duty along the East Coast from Maine to Key West, Florida as part of Destroyer Division 8. The young sailor was discharged in May 1947, receiving numerous medals and ribbons. In 1950, he married his wife Nancy to whom he has been married for fifty-nine years. Coming home, he was treated like a hero. From his enlistment until today, he has been honored and respected for his service to his country to which he is ever thankful.

After discharge, he went to the Baltimore Business College under the GI Bill. He was hired by the Sinclair Refining Company in 1951 where he worked until 1981. In 1982, he again joined the work force and works at his present employer, Griffith Oil Company.

The greatest event in his life was when he was twenty-one years old. He heard the Gospel given and accepted Christ, becoming a Born-Again Christian. John says to this day that any success he has had, "To God be the Glory." His wife, Nancy, has been diagnosed with "end stage" Alzheimer's disease for the last four years. Since this writing, she has gone to the Lord.

He lives today in Westminster, attends Church of the Open Door, and enjoys life with his three boys, eight grandchildren, and eleven great grandchildren. This sailor served his country well and deserves the respect and thanks of all who meet him.

Hayes Larkins
1st Lieutenant
7th Infantry Division, 17th Infantry Regiment, L Company
Korean and Vietnam Wars

November 26, 1951…Second Lieutenant Hayes Larkins and his patrol of 35 men moved forward toward enemy lines. Heavy snow fell as they approached, seeing no activity. He sent out 2 lead scouts to survey the forward line. The scouts were engaged by North Koreans. That was when all hell broke loose.

Hayes was born March 22, 1929 in Washington, D.C to Carlton and Elizabeth Larkins of Alabama. His father was a career Navy man and was assigned to the US Navy Yard in Washington, D.C. at the time. As an infant his Mother later carried him to Port-au-Prince, Haiti to join his Father who was assigned to the USMC expeditionary unit there. His sister was born there one year later. His Dad had enlisted in the Navy in 1922 and served for thirty and a half years reaching the rank of Lieutenant Commander. A highlight of his younger years was his appointment as a Page on the Democratic Bench of the US House of Representatives, from February to September 1945. In this position he was on the scene of a number of significant events. One such event was the joint session of Congress being addressed by President Franklin D. Roosevelt following his return from the Yalta Conference. Because of his father's various Navy assignments Larkins had changed schools ten times before he was graduated from the High School Division of the Marion Military Institute (MMI) in Marion, Alabama in January 1946 at age 16.

On April 30, 1946, following his 17th birthday Larkins enlisted for a short-term (18-month) in the Army Air Corps at Baltimore, Maryland. He immediately took basic training at Amarillo Army Air Base in Texas. After assignment in Boca Raton, Florida, he went to an Army Overseas Depot at Kearns, Utah. From there he went to Chitosi, Japan on Hokkaido. Larkins then went to Surgical Technician training at the 28th General Hospital in Osaka, Japan. He was transferred to the U.S. Military Academy Prep School at Stewart Field, New York for three months in training and eventually was discharged October 30, 1947.

At age 18, Larkins enrolled at George Washington University in Washington, D.C. on the G.I. Bill of Rights. Not having done well academically at GWU, he went back to MMI and was graduated May 25, 1950 with General Academic Honors and an AA degree. In addition, he was commissioned as a 2d Lt, Infantry, in the ORC of the US Army.

On June 20, 1950 he reported to Fort Benning, Georgia for 90 days of active-duty training for the Infantry Associate Officers Course. Five days later he was informed that the North Koreans had crossed the 38th Parallel. He wondered "where in the Hell is the 38th Parallel". He didn't realize that in less than one year he too would be making the crossing. He completed the course on September 25th and left active duty from Fort Benning.

Larkins reported back to active duty at Fort Benning on October 3, 1950. Following assignments with the 78th Combat Engineer Battalion and with the 15th Infantry Regiment he was ordered to Korea in May 1951. He was assigned to the 7th Infantry Division, 17th Infantry Regiment, L Company where all its members were "requested" by the Regimental Commander, Col. William "Buffalo Bill" Quinn (later retired as a LT General) to grow moustaches. Therefore, they were known as "Buffalo Stashes." While on emergency leave, his unit engaged in some of the heaviest fighting at Chup'a-ri. They endured heavy casualties. Upon his return from leave he was assigned to I Company.

On November 26, 1951, his patrol edged forward as two lead scouts encountered North Koreans. Madness began as heavy fighting broke out. As his men huddled in the ice and snow, allied artillery finally began pouring in to help them. It was so chaotic, that as they finally broke out to get away, their own troops began shooting at them thinking they were the enemy. Larkins had been hit in the arm and foot by enemy fire. Of his 35-man patrol, all but 11 were either killed or wounded. He was evacuated to Japan to the same hospital where he learned to be a surgical technician. While there he was promoted to 1st Lieutenant.

He returned to Korea in January 1952 to a weapons platoon. One night they watched as an endless column of enemy troops moved to the left and one also to the right. They were told to have no movement during the day and no lights at night (this was during "Operation Clam-Up"). As darkness enveloped their position, they thought they saw movement in front of them and asked permission to shoot and were denied. They asked again and denied. He told a South Koreans soldier to throw a rock forward to see if anything happened. A flare burst overhead, and everyone began firing. They found out later that no one had been out there! In March, he returned to an assignment with the 47thID at Camp Rucker, Alabama. There he had various assignments as a company commander, company executive officer, and battalion staff officer. His last assignment was as an instructor at the Army Aviation School.

In 1956 the Army began its "Reduction in Force Program (RIFF)." He was "riffed" in March 1957 and enlisted as a Sergeant First Class in the Army's Counter Intelligence Corps (CIC). After training as a special agent at Fort Holabird, Maryland, he spent a year working in Denver, Colorado and then a year at the CIC Field Office at Fort Rucker, Alabama. In 1959 he was chosen to attend the Army's College Program that had begun in 1958. After matriculating for two years at Michigan State University

(MSU) in East Lansing, Michigan, he received his BS degree with High Honor in March 1961 and completed a year of graduate course work. In August 1961 he received orders to report to the 513th Intelligence Group in Germany due to the Berlin Crisis.

He spent three years with the 513th working first in the Counter-espionage Unit and then in the Unit's Security Office. This office had the responsibility for the internal security of the Group and all of its personnel.

In August 1964 he reported to the United States Army Intelligence School where he was assigned as an Instructor. This assignment lasted for three yearsand during this time period he completed his Thesis and was awarded a master's degree from MSU.

Larkins received orders in July 1967 for assignment to the 4th Military Intelligence Detachment, 4th Division in Vietnam. Following his arrival in Vietnam he operated as an Intelligence Agent in the Central Highlands with Montagnards for two months and then moved to Duc Pho where he supported the 3rd Brigade of the 4th ID. During a five-month period he conducted field operations of a classified nature with elements of both the 4th ID and the Americal Division making 14 Air Combat Assaults with these units. For these operations he was award the Vietnamese Gallantry Cross. Following the Tet Offensive, he was assigned to the Headquarters of the newly formed MACV Forward Corps Headquarters in Hue-Phu Bai.

Larkins returned to the U.S. in August 1968 and again was assigned as an instructor at the Army Intelligence School and retired in January 1969 having served 20 years of active duty and achieving the rank of Captain, USAR. He received 17 medals including his coveted Combat Infantry Badge. He is also proud of his Purple Heart and Vietnamese Gallantry Cross with Bronze Star.

He became Assistant Professor of Criminal Justice at the Community College of Baltimore, retiring as Full Professor after 25 years. From 1974 to 1980 he taught part time at the University of Maryland College Park while working on his Doctorate.

Larkins has lived in Hampstead, Maryland since 1970 and has been retired from the College since 1994. Following his retirement, he has traveled extensively in Europe, Central America, and China. He has also participated in the "Return to Korea" program which is sponsored by the Ministry of Patriots and Veterans Affairs and the Republic of Korea. He made the return trip to Korea in June 2010. He has been married for 58 years to Eudora Carmichael of Enterprise, Alabama and has one son and four daughters. In all, he has 8 grandchildren and 1 great granddaughter.

Charles Franklin LePage
Lieutenant J.G., U.S. Navy
Hellcat Fighter Pilot, Fighting Squadron 1
Aircraft Carrier U.S.S. Bennington CV 20
World War II

All of the training was now over. After two years, the Hellcat F6F fighter pilot, Frank LePage, was finally on his first mission. The sixteen aircraft of Fighting Squadron 1 of the Aircraft Carrier U.S.S. Bennington, CV-20, were heading for Japan…payback time. Suddenly, radio chatter broke up the suspense as a "bogey" (unidentified aircraft) was picked up at ten o'clock high. The young fighter pilot felt his resolve begin to diminish as a new fear began to overcome him. His eyes welled up as the thought of combat became overwhelmingly real. Chatter soon confirmed that it was a "friendly." LePage knew now in his heart that this was real…this was war.

West Virginia parents Frank and Nellie Le Page bore their son Charles Franklin LePage on October 11, 1922 where he would live with his two brothers and sister. He started out as a control chemist for a gas distribution company and eventually ended up in Morris Harvey College in West Virginia. On October 13, 1942 he decided to do his duty and enlisted in the U.S. Navy and was assigned to the Navy V-5 program to study becoming an aviation pilot. With two years of training behind him, LePage was stationed at Fallon, Nevada on the Navy Auxiliary Base where he joined Fighting Squadron 1 that was organized in October 1944. Further training came on the Island of Maui in Hawaii in January 1945 for three months. LePage talks of spending more time on railroad trains than ships from all the training he received.

Assignments came and the squadron was stationed on board the Bennington. The massive carrier just came through the monstrous typhoon that had struck Okinawa and caused significant damage to the fleet: one of them being the Bennington. Damage included the catapults that launched aircraft. LePage was not on board during the typhoon.

Being a fighter pilot required extreme concentration. Aircraft were landing usually at 45 knots on a rolling vessel. They would fly parallel to the carrier, do a 180º downwind, and do another 180º into the stern of the ship. A Landing Signal Officer on the edge of the deck would pick them up at 90º from the landing position and direct them in as "OK," "slow down" or "speed up." If "OK," the officer would flash a wand across his neck signaling the pilot it was clear to land. With the tail hook on the aircraft extended, the pilot would cut power and fly toward the deck. Nine steel cables traversed the deck. The tail hook would grab one of them and the craft would come to a very abrupt stop. LePage's first landing was in training at Great Lakes Naval Recruit Training Center and noted it was not that difficult. By the time he arrived on the 872' long Bennington, it was "old hat."

Part of LePage's duties as a fighter pilot included aerial reconnaissance. His plane would fly over targets before and after to access targets and damage. During his active combat, he had ten contacts with the enemy, primarily taking aerial reconnaissance of enemy targets, of which he received several commendations.

From July 10 to July 30, 1945, he flew several successful photographic missions over the Japanese homeland. The excellence of his photographs materially contributed to damage assessment and military intelligence in the planning of subsequent operations against the enemy.

On July 28, 1945, during a combined photo and radar mission preceding a coordinated task force attack on enemy Japanese warships at the Kure Naval Base on Honshu, Japan, he made two runs through intense and accurate antiaircraft fire from warships and shore installations. He employed special radar equipment which effectively deflected the fire from the main attack group. LePage successfully photographed the targets assigned the main task force and then returned to base despite hazardous weather.

The young pilot received the Air Medal for action on August 13, 1945. During action against the enemy off the coast of Honshu, Japan he targeted an enemy Kamikaze Betty Bomber making a run for one of the destroyers of the task force. LePage attacked the hostile aircraft, firing his six fifty calibers machine guns, and shot it down in flames into the sea.

Finally, the cease fire came on August 15, 1945. The war was over. The Bennington was at the formal surrender of Japan with her aircraft joining others as they flew over the ceremony in an impressive display of victory.

Some of the Bennington's planes, including LePage's, continued taking photo reconnaissance of the Japanese Islands. They even dropped candy and cigarettes over P.O.W. camps to waving hands and hoping none of the boxes would hit the P.O.W.'s. For his service, he received 2 Air Medals, the Distinguished Flying Cross, Navy Gold Wings, the Pacific Theatre Ribbon and Good Conduct Medal, among others. He was discharged in December 1945.

After the war, he went back to Morris Harvey College and finished his degree, eventually working there. He received his master's at Ohio State University. Various other jobs would be at the Probation and Parole Board of West Virginia and also as a volunteer of the Police Civil Service Commission for eight years. He eventually retired from State service in 1988. He has dabbled in real estate agencies part time since then. He has since retired and moved to Sykesville, Maryland in 2001 where he lives with his wife Nancy of sixty-five years. LePage has one son and one daughter and five grandchildren. This fighter pilot has done his duty to his country and served it with honor.

Charles Leppo
Sergeant
B Battery, 304th Field Artillery, 77th Division
U. S. Army
World War II

Once the shelling had stopped in the Battle of Okinawa, Sergeant Charles Leppo's 105mm Howitzer field Artillery piece rested. His crew relaxed in the lull. Their captain told them to get in the trenches that the Japanese forces had earlier made. The enemy shell came from behind, leaving no noise of its approaching trajectory. The missile plowed into the seven-foot-long trench where Leppo rested, burying him alive in three foot of dirt. The gun crew frantically dug him out. He was alive. As he was being pulled out, his eyes saw the large, round projectile, fired from an enemy ship, lying in the dirt next to him. It had miraculously been a dud and never exploded.

Born in the Smallwood community of Carroll County, Maryland on September 29, 1920, Leppo and his five brothers and two sisters worked mostly on farms. His parents, Claude and Estella, kept them busy. Upset because of Pearl Harbor, like so many other young Americans, the young patriot enlisted in the Army in January 1942. Little did he know that of the four years of military service he would spend thirty-six months of that time in actual combat.

Leppo took his basic training at Fort Bragg, North Carolina and later transferred to Fort Jackson, South Carolina and was assigned to the 77th Division, a former World War I division reactivated. They again went back to Fort Bragg for more training and field service training on all types of terrain. Desert maneuvers were next in Arizona. They had to be prepared for anywhere in the world.

As part of the Field Artillery, Leppo was assigned to 105 mm Howitzers. These guns could be fired two to three times a minute, depending on the crew, and up to eight miles. Each shell could take up to eight powder charges.

When they were finished in Arizona, the soldiers headed for Camp Stonewood, California, near San Francisco. Then it was off to Hawaii and the Pacific Ocean. He arrived in Pearl to see some ships still burning from the attack of December 7th. Training continued for the recruits for some time. The army finally put them to the test. The Eastern Mandates, composed of the Marshall and Gilbert Island chains, was on the list.

Guam, itself, lasted only two to three weeks. The 77th had a difficult landing. Lacking amphibious vehicles, the men had to wade ashore from a nearby reef where they were dropped off by landing craft. As in most Pacific battles, the Japanese defenders refused to surrender, leaving most of them dead.

He then remembered Mindanao which was more or less getting themselves together for combat. Cebu Island lasted only a week.

The Philippines saw some of the most intense fighting by the Japanese. Leppo was involved in three battles on the islands. The soldier saw many casualties but, luckily for them, there were not many in the Field Artillery. Okinawa was where Leppo saw a lot of action with his artillery unit. He still is amazed how many of the islands were honeycombed with tunnels and caves, hiding the enemy. The 77th also helped in the liberation of the Ryukyu Islands.

Leppo recalls one of his best officers, Captain George Bowers. Bowers was a criminal lawyer at the age of twenty-four. He was never too busy to talk to the enlisted men. To Leppo, that meant a lot. Bowers made it through the war also and was transferred to Corps Headquarters.

When the war ended, the 77th became an occupation force. Young Leppo was stationed in Yokohama, Tokyo,

Hakodate, and Sapporo and a handful of other small towns. After the war, he was sent to Seattle, Washington. He was discharged December 31, 1945.

During his time, he was fortunate to see several USO shows. Some included entertainer Bob Hope and many others. Some shows were stateside and some overseas. Of the war, he says he doesn't regret one day of it. He did what he had to do. One thing he quotes that made combat interesting was that everyone was treated the same: there was no difference, even in all the services.

After the war, he worked for the Baltimore Gas and Electric Company for thirty-four and a half years before retiring.

Leppo has attended 77th reunions every year for several years since the war. The last one he attended, out of the 365 in his unit, only 27 are left. Several members of Leppo's family also served in the war. In Hampstead, Maryland, where Leppo now lives, is a War Memorial. He helped to build it and it, too, bears his name upon it. He has also been an active member of the American Legion for over sixty years, now attending Post 200 in Hampstead.

Sergeant Leppo lives in Hampstead with his wife of sixty-five years, LuRay Leppo, whom he married in 1945. He has three children, one who lives in Nevada, and seven grandchildren.

John Looser
Master Sergeant
1st Anti-Submarine Squadron
Army Air Corps
World War II

At Selfridge Field in Mt. Clemens, Michigan on a quiet Sunday morning, young John Looser sat in the Code Room typing any messages that came over WAR, the War Dept Radio Network. Some messages were minor in nature and some were to be sent immediately to their designated parties. Suddenly, one really got his attention. He began typing the urgent message and began to sense a feeling of anxiety such as he had never felt before. The message began something like this: "Sunday, December 7, 1941. American forces in Honolulu, Hawaii have been attacked by Japanese armed forces."

Looser was born February 14, 1921 in Holcombe, Wisconsin to Ernst and Ida Looser. His father was an immigrant from Zurich, Switzerland. He had three sisters. During high school he worked for the Wisconsin Cooperative Milk Pool. After graduation from High School, Looser worked in a cheese factory with his Dad.

John Looser entered the service on October 9, 1939. He was to train for control tower operator for the Civil Air Authority (CAA). At that time all CAA tower operators were trained by the Signal Corps and released when training was completed. He went to boot camp recruit training at Selfridge Field in Mt. Clemens, Michigan. Then it was off to Chanute Field in Illinois for a Control Tower Operators course, followed by radio operator, mechanic, and teletype schools. While there, they hauled away thousands of cylinders used for helium for the dirigibles. The release program ended before Mr. Looser completed his training.

Looser then went back to Selfridge and on to Clinton, Ontario to attend a Royal Air Force (RAF) radar school. Again, he went back to Selfridge and became part of the 5th Fighter Squadron, a P-39 unit. In early 1942, at MIT the US Army Airs Corps formed the 1st Anti-Submarine Squadron at Logan Airport in Boston. He helped make ten radar systems for ten B-18 Bombers. They then began patrolling from New York to Trinidad searching for submarines. Looser says they never sank any but turned the info over to the navy who did sink many of the enemy subs.

In 1942 he recalls flying in a B-18 near Key West, Florida. He was the radio/radar operator. They tried landing in a storm with no navigational aids. They crashed on a coral reef and tried to cut the power switch, smelling fuel. He jumped out, cutting his legs badly on the coral but they all walked away.

Eventually they turned the B-18's in and got new B-24's which they were to fly over to England. They lost one plane due to an ice storm. In England they joined the British 1st Group Coastal Command at St. Eval. Looser, now a Radar Maintenance Chief then went to Morocco, near Casablanca, where they flew missions over the Atlantic looking for sub packs. Later they flew from Tunisia over the Mediterranean doing the same. The unit was then abandoned and Looser went to Air Force Test and Evaluation Center at Orlando, a weapons Research and Development site. He received his discharge in 1945.

He worked at Kohler of Kohler and attended night school taking math classes. He missed registration dates at the U of Wisconsin and Tri State College and decided he would do better back in the service. He re-entered the Air Force in 1946 as a Master Sergeant. He was assigned back at Selfridge in the 56th Fighter Group and became crew chief of the F-86 Fighter.

As part of the 56th at Selfridge he was assigned to fly with the 56th Fighter Group making the first Atlantic crossing with fighter jet aircraft. He would be part of the ground maintenance crew and maintain the radio equipment and work on the aircraft electrical systems. The ground crew, spare parts and ground handling equipment would follow in a C-54 cargo aircraft. The crews flew from Selfridge to Bangor, Maine, then to Labrador, Iceland, Scotland, and on to England.

When they landed in England, the crew chief of the C-54 told the British service crew to "fill her up" and left. The plane had two large tanks in the cargo section for extended range and they were empty for their trip as they were at maximum load. By refilling the large tanks, they craft became overweight and on takeoff they were running out of runway. The pilot got the plane up just as it clipped telephone wires. A large cliff waited for them as they began to drop down to the ocean while trying to pick up speed. The plane slowly began to pick up speed and they flew on to Germany. When they got to Furstenfeldbruck, they had to circle for five hours to burn off fuel and get to landing weight. The crew chief was taken to jail for court martial and a new crew chief was assigned.

BOMARC Missile site at McGuire Air Force Base was next after he left the 56th and part of the 46th Missile Squadron. He then left the 46th to go to IBM School in Kingston, New York and back to McGuire Air Defense Command as Chief of Maintenance for the computer Center. This was during the Cuban Missile Crisis. He recalls operating 24/7 with no down time.

In June 1960, while at McGuire an IM-99 Missile exploded and caught fire. Looser dumped all the helium and depressurized all the missiles. They called in nuclear experts in and saw no danger from radiation. They would find out later the experts did find danger. Because he deactivated all 45 missiles, he was court-martialed but later found to not be at fault and reinstated. It wound up being a flaw in the design of the missile helium tank. Looser then went to Germany in

1963 to work on the 412L Air Defense System and eventually retired October 6, 1965 after twenty-seven and a half years of service.

After his service, he worked for Bendix Corporation for 1 year working on the NASA Tracking Network. He then went to Goddard Space Flight Center for 13 years as head of the network support team assisting on programs such as Gemini, Apollo, Skylab and Soyuz. He then retired to Hanover, Pennsylvania but continued helping on the Board of Directors of Adams County and Allegheny Electric Cooperatives for 12 years before moving to Eldersburg, Maryland in 2006.

During his military service Mr. Looser held every grade from private thru major including the four Warrant Officer grades.

Some of the Air Force awards received by Mr. Looser were THE LEGION OF MERIT, THE OUTSTANDING PERFORMANCE AWARD with TWO CLUSTERS. NASA awarded him with the Snoopy and OUTSTANDING PERFORMANCE Awards for his work on the Manned Space Flight Missions. The Cooperatives in PA voted him THE OUTSTANDING DIRECTOR OF THE YEAR.

Today he lives at Fairhaven Retirement Community in Sykesville, Maryland and enjoys seeing his 2 sons and 2 daughters and 11 grandchildren. His wife, Jean, passed away in 2007. They were married 61 years.

James MacDermott
Staff Sergeant, Ball Turret Gunner B-17
8th Air Force, 388th Bomb Group,
563 Bomb Wing
Army Air Force
World War II

The bombing group headed for its target. Young twenty-year-old James MacDermott hunched in the seat of the belly of the mighty B-17 Flying Fortress's ball turret. His eyes scanned for the enemy aircraft. Flak from below began pummeling the air space around the group of aircraft ahead of MacDermott's. One by one they began taking hits. One craft would go down and then another. Eight craft were lost from the group. As MacDermott's group neared the flak zone, it suddenly stopped. As they passed through the danger zone, the calm was broken up by the flak again as it pounded the air space behind them. To him, it was a miracle.

James MacDermott was born in Baltimore, Maryland to James and Leotta MacDermott on November 18, 1925. While in Forest Park High School, recruiters came to the school promising the young men they would get whatever assignments they wanted. MacDermott wanted Aviation Cadet School. He wanted to fly. The young man never got it. He left school and enlisted on March 18, 1944 in the Army Air Force.

The military wasted no time in sending MacDermott right off to Keisler Field in Biloxi, Mississippi for boot camp. Then he was off to Kingman, Arizona for Flexible Gunnery School training. Dyersburg, Tennessee was next for Combat Crew Training School. Several months passed and MacDermott then went to Lincoln, Nebraska with a crew to pick up a plane to go overseas. However, something happened that changed things dramatically.

December 16, 1944 the Battle of the Bulge began. MacDermott's crew getting their own craft was put on hold. They were instead sent overseas to Hopton, England, a former RAF airbase. Even though they were assigned a B-17, they still used several others. Their pilot, on his own, named their craft the "Sleep Robber 008." It pictured a vivacious female sitting on a falling bomb.

He recalls landing in Belgium once at night due to a fuel problem. By noon the next day they were back in the air and flew back to their base. The young ball turret gunner recalls how it took a while for planes taking off to get into formation. Sometimes there were up to three hundred planes and as they took off one by one, they would form a large circle in the sky until they were all up and ready to head off on their mission. But the farther the ground troops got into Germany, the shorter the time lining up.

The Fortress was no stranger to protection and weapons. They sported two .50 caliber machine guns in the nose, two behind the cockpit, two in the ball turret, two in the tail and a single .50 caliber on either side of the nose and either side of the fuselage. They also carried the Norden Bombsight, a gyro-stabilizer computer that was used mainly during daylight bombing missions. It was basically a point and shoot device that help improve accuracy. Losses from flak continued to take a heavy toll on the B-17's through 1944.

MacDermott and a friend signed up together when they entered the Army Air Force but decided not to stay together. As a result, his friend, on another mission, lost an eye and got shot up pretty bad.

He talks about the Togalier's on the B-17's nearing the end of the war. Togaliers were "dumbed down" bombardier's who would wind up being a chin turret gunner. When the mission did not require the services of all bombardiers on every plane in the formations, they would put those men in the chin turrets. Only the lead plane and the next in line would use a bombardier. But when they neared their targets, a flare would be dropped for all the craft in formation to watch the lead B-17 open its bomb bay doors and begin dropping their bombs. The others would then follow suit.

Upon arrival back at base, the crews were always given a shot of whiskey before their briefing. The briefings were mainly for the officers and pilots. He states that if they wanted to know where they were going, they would listen to Axis Sally on the radio.

After their 35 missions and the same crew throughout, they went to Stone, England to wait for a ship to go home. He arrived in New York and went to Fort Dix on a pass and got to go home for ten days. They were on a train heading to Florida to be reassigned to B-29's when they learned of VJ day. The war was over. MacDermott was discharged November 6, 1945. He received several medals and ribbons including three bronze stars and an air medal with a silver cluster.

He came home and went to work for Adler Display for 25 years and Runkle Signs for 5 years before retiring at the age of 65 and then drove for CATS senior Bus service for 10 years. He and his wife of 53 years moved to Carroll County in the 1960's. He has 1 sister, 1 son, and 1 daughter including 1 grandchild.

Charles Magee
Photographers Mate 1st Class
United States Navy
World War II

This story begins for a newborn on March 20, 1921. His name was Charles Magee. Charles was born to John and Hilda Magee. The family lived on a working farm in Reese, Carroll County. He was raised and worked the farm with his family.

Charles ended his high school education in 1938 when he graduated from Westminster High School. He then started what would be his lifelong ambition. In 1938 he began working at the Westminster Photography Studio. The amateur photographer wanted to learn all he could about the art.

When WW II broke out on December 7, 1941, Charles did not waste any time. He, like so many other young men, heeded his nations call and enlisted in the military. On March 21, 1942 he joined the U.S. Navy.

After Boot Camp, he was assigned to the destroyer U.S.S. Overton. As a Seaman, his duties included "swabbing the decks" which he hated. He would do anything to get out of that duty. So when some crewmen convinced him he should go into radio work, he became a "radio striker." A striker is someone who is trying to learn the position from those already in the position. He says he never really cared for it but it got him out of the other duties. But fate had its alternatives.

A memo came around stating that the Navy was looking for people interested in photography. It was like a Godsend to Charles. He put a request in for the school. He never got an answer as quick as he thought he should when he found out that the Executive Officer was tearing them up. After three months on the destroyer, he was finally able to convince the officers to let him go.

Charles went to Photography School in Pensacola, Florida. He was taught in basic, advanced, aerial photography, and movies. The classes were one month each. He was then assigned to the Atlantic Fleet Camera Party. Their main purpose was fleet photo triangulation. They would take photos at different locations of ships firing ammunition and make sure it was going to the right targets. According to Charles, through the photos, they would be able to recalculate the ships computers and make sure they were firing correctly and precisely. He did this on all kinds of Navy ships including Carriers, Battleships, Cruisers, etc.

On one occasion, Charles remembers that after they were left on an island to do photos for a firing exercise on Bloodsworth Island in Chesapeake Bay, he and three other personnel were forgotten and left on the island after the exercise. Eventually they remembered them and picked them up.

One moment he is very proud to remember is that while in Cuba, he sang in the Navy Choir and also in a quartet with a very famous person who was in the Navy at that time also. His name was Archie Campbell who would be a star on the TV show "Hee Haw."

Some of the stations he served at were Casco Bay in Portland, Maine, Norfolk, Virginia, Guantanamo Bay, Cuba, and Trinidad. Magee rose to the rank of Photographers Mate 1st Class. He was discharged from the Navy in October 1945. He stated that to be precise he served three years, eight months, and fourteen days!

After the war, Charles returned home to Westminster, Maryland and bought the studio that he had previously worked for. That is where he had worked for forty years before retiring in 1985. He also was a master at woodworking and, over the years, has made some beautiful furniture, some of which is in his current home.

In 1944, he married his high school sweetheart, Madge Brown. They have been married for sixty-five years and have two children, John and Bonnie. Time today is spent with their five grandchildren and five great grandchildren.

About his time in the Navy, Magee says "he met a lot of good people." Though he was not on the front lines, his service was still valuable in the victory our nation achieved in the Second World War. Everyone has to do their part for the good of the whole. Charles certainly did his. Since the initial writing of his story, Charles has departed this world.

Russell Mayer
Private First Class
508th Parachute Regiment, 2nd Battalion
Medics
World War II

Born in 1923, Russell Mayer was raised by his mother, Bertha Mayer, in the Pimlico area of Baltimore City along with his sister and five brothers. As he grew, he became an entomologist and even worked at an A & P grocery store in Pikesville. He also volunteered for the Pikesville Volunteer Fire Department as a medic.

As WW II began to turn in favor of the Allies, Russell was drafted on January 14, 1943. The twenty-year-old began a trek that would take him to places he would have never imagined. He went to Camp Blanding, Florida where he saw paratroopers waiting for them. They were forming the 508th Parachute Infantry Regiment. Asked if he wanted to be a paratrooper, he firmly said no. They thought that being a volunteer fireman he would also like to jump out of planes. They said they had other jobs for him.

After basic, he was assigned as a cook. After receiving an injury, he was laid up for a while. Upon return, he was sent back to the 508th. In late December 1943, the regiment was loaded up in New York on the *U.S. Army Transport James Parker* and sent to Belfast and then ferried to Nottingham, England. In June '44, they moved to Saltby Air Base to prepare for their first combat jump on D-Day, June 6. The regiment would jump near the towns of Ste. Mere Eglise and Picauville, west of Omaha and Utah Beaches. The rest of the regiment, including Russell, went across the Channel on boats.

Mayer caught up with his regiment on June 7 and acted as an infantryman and a cook. He still vividly recalls the rows of covered bodies lined up on the beach as they came ashore. As they went inland across a field, they saw a glider that had crashed nose down into trees and the bodies of the troops still in it. On July 13, they prepared for their next mission. Of the 2,056 men of the regiment who jumped into Normandy, 1,161 became casualties.

A captain from the Medics told him he was no longer a cook but a medic because of his previous training on an ambulance. On September 17, 1944, they jumped into Holland and captured the Nijmegen Bridge. Russell recalls as he floated down that he could see the slip trenches and thought for sure that Germans were ready for them, but they received very little resistance on the drop. The bridge was the prize of the operation.

During the first few days in Holland, British troops were winding their way north of Belgium to reinforce British and Polish troops who were trying to capture a bridge at Arnhem. But at 3 PM in the afternoon, the British troops stopped and had tea and crumpets! Then they got back in their tanks and continued on. But it was too late, the British and Polish in Arnhem had been decimated.

On December 16, 1944, they were loaded on trucks and sent to Belgium where they would become involved in the Battle of the Bulge. On January 9, 1945 while still in the Bulge, he was ordered to get the injured together for transport. After returning to the Battalion Aid Station, his captain checked him out thinking he was getting a cold. Russell's feet were tingling so he loosened his boots. The Captain told him to take them off and realized that Russell's feet were frozen...frostbite. After a few months in Liverpool, they sent him back to the States on the same ship that took him over, now converted into a hospital ship. On the crossing, they found out that Germany had surrendered. When he debarked, on Mother's Day, he kissed the ground. Russell was discharged August 8, 1945.

Russell vividly remembers many of the sights and horrors he saw which will not be put in this story. Often times his chin still quivers and his head will go down as he retells some of the events. For his service, he received fourteen ribbons and medals, including three from the French and Dutch.

In June 1948, he married Zelda Grey Holcomb, his wife of 61 years. They have a son and daughter, 5 grandchildren and 7 great grandchildren. After his service, he became a Baltimore County Police Sergeant and served for twenty years. Today, Russell lives at the Fairhaven Retirement Community in Sykesville, Maryland with his Zelda.

William J. McKenzie, Corporal
3rd Army, 80th Division
319th Infantry Regiment, 2nd Battalion, H "Co."
U.S. Army
World War II

The morning of December 23, 1944 after they took over the town of Heiderscheid, Luxembourg, Corporal William McKenzie looked out the second story window of the farmhouse and saw men camouflaged in white sneaking along the road behind the house. He knew the Allies didn't wear white, so he fired at them, wakening the rest of the Americans. After about ten minutes of firing, 29 German tanks and half-tracks loaded with German soldiers came up out of the draw and surrounded the town: 5,000 Germans against 263 Americans. Young twenty-year-old William McKenzie, after hearing about the recent massacre at Malmady, figured he was finished.

McKenzie was born in St. Clair, Pennsylvania to George and Mary McKenzie on May 31, 1922. After graduating from St. Clair High School, he worked for the National Youth Administration and learned the trade of acetylene welding. This allowed him to go to work for Oriole Gas Stoves in Baltimore, Maryland. While living in Baltimore, he received his "Greetings from Uncle Sam" letter telling him that he had been drafted. He registered in June 1942 and was drafted on October 17, 1942.

After induction at Fort Meade, Maryland he attended boot camp at Camp Wheeler, Georgia. He went back to Fort Meade where he was assigned to the 76th Infantry. The young soldier then went to Camp McCoy, Wisconsin. The 76th went to the upper peninsula of Michigan for winter maneuvers. He was then sent to Fort Dix, New Jersey as a replacement in the 80th Division to be a messenger/runner. Finally, he boarded a large ship which wound up being the converted ocean liner Queen Mary.

They arrived in Glascow, Scotland and ferried on train to Southampton, England. They then boarded an English ship and sailed across the channel to Utah Beach, landing on August 3, 1944. While the Germans were still shelling the beaches, they headed inland to the front lines. The battle of the hedgerows had been ongoing since D-Day. General Omar Bradley told General George Patton that he was tired of being bogged down and wanted an old tank man to make an end run. The 80th was ready.

They moved like lightning across France to the Seille River, then the Saar Basin and eventually to the Siegfried Line. Once the Germans started the Ardennes Offensive, they moved into Luxembourg fighting the Germans southeast of Bastogne. He recalls walking 30 miles one day and 32 the next.

Once the Battle of the bulge started, they drove in trucks all night from Saarbruken to the city of Luxembourg, 180 miles away. The 80th moved toward the 101st Airborne trapped in Bastogne and made a night attack on December 21 on the town of Heiderscheid. They took over the town. The next day as he guarded the Command Post a German truck came down a road toward the town. He shot the driver and the truck crashed. The truck had been loaded with German troops. After a small skirmish, they killed all the enemy troops. This happened right in front of battalion headquarters!

As daylight broke on the 23rd, McKenzie looked out the second story window of the farmhouse and saw men camouflaged in white sneaking along the road behind the house. He knew the Allies didn't wear white, so he fired at them, wakening the rest of the Americans. 29 German tanks and half-tracks loaded with German soldiers came up out of the draw and surrounded the town. Using bazookas and rifle grenades the Americans set 11 tanks and half-tracks on fire, burning to death many of their occupants. The other 18 tanks and support vehicles withdrew and moved on to Bastogne.

On December 29, the enemy launched a counterattack at Ringel, Luxembourg threatening McKenzie's command post. He held his position at the command post, killing 3 of the enemy, wounding 5 and forced the remainder to find shelter in a barn. He then ran to the barn and forced what he thought were 2 but wound up being 16 of the enemy to surrender. Later when he observed an enemy rocket launcher team in firing position he crawled within range to throw a grenade, repulsing the attack by killing 2 of the 3 of the team and wounding another. For this he received the Distinguished Service Cross.

McKenzie and a few soldiers rested in a farmhouse in Dahl, Germany, feet against the fire and sitting on a sofa. They could hear shells being fired over them. They could almost tell by the sound of the shell if it was going to go over, hit them, or fall short. All of a sudden, a shell hit the far wall of the house. He recalls only seeing a flash. A lieutenant found him covered in debris and on the opposite side of the room from where he was. He survived with only dent the size of a baseball in his helmet. One man suffered a kneecap missing and part of his left arm while another took shrapnel in the groin.

On April 4, 1945, he was summoned all the way up from Command Post (CP) to Battalion CP, to Regimental CP, Division CP and Corp CP for something he knew nothing about. He thought he was in big trouble. It was a flag raising ceremony for Army Day. General's Patton, Bradley and Montgomery shook hands with him. The most decorated soldiers from each outfit in the 3rd Army were invited. It was a day he would always remember shaking the hand of George Patton.

One of the rewarding but disturbing events that McKenzie participated in was the liberation of 4 Polish concentration camps, most notably Buchenwald and Mittelbau Dora. He like so many others says that those who say the Holocaust never happened are liars. The 80th went all the way to Austria and waited for the Russians. For the remainder of the war, he was a jeep driver and captain's orderly.

McKenzie was discharged October 21, 1945. He married Wilma McKenzie, his wife of 63 years, on June 29, 1947. He returned home to work for Oriole, then Locke Insulator and finally the Social Security Administration until he retired in 1977. He had 7 brothers, five of whom served during the war and 1 sister. They moved to Carroll County in 1971. He enjoys his hobbies of model trains and collecting model cars and a real 1948 Dodge. He also is a life member of the Reese Volunteer Fire Company and the American Legion Post 31. He is one of the most decorated soldiers in Carroll County.

Vernon Miller, Sergeant First Class
2nd Infantry Division, Combat Engineers
B Company
U.S. Army
Korean War

The Combat Engineers of the 2nd Infantry Division were told to help a French Battalion fortify Arrowhead Ridge. Sergeant Vernon Miller and his men assisted them by putting in mine fields, barbed wire with trip explosives, napalm and drums with detonating cords. But their main job was to make sure the pontoon bridges weren't damaged. They had completed most of the job at night. As morning light came, mortar rounds started falling all around them. Then they heard it…thousands of Chinese troops were racing and screaming up the hill toward them.

Vernon Miller was born and raised in Carroll County, Maryland on December 2, 1929 to Russell and Hilda Miller of Westminster. Miller graduated from Westminster High School in 1947 and worked for his dad at Miller's Electric Service, a family business.

With the Korean War in progress, he found himself drafted on July 6, 1951. Miller was processed at Fort Meade, Maryland and sent off to Fort Belvoir, Virginia for recruit training where he finished up with Electrician School. He remained there until Christmas and furloughed over into 1952. While in school, he is proud to say he was picked as "Soldier of the Week."

Next was Seattle, Washington where he boarded a liberty ship heading for Japan. He arrived at Camp Drake in Yokohama for training and remembers how beautiful Mt. Fuji looked. Miller took a train to Sasebo, going through Hiroshima recalling how desolate it was. He was then picked for schooling at Gifu, an old Japanese Zero base. After training, it was back to Sasebo and onto Pusan, Korea.

He was assigned to the 2nd Division Combat Engineers, B Company, and while settling in he could hear artillery off in the distance. His sergeant told him to get used to that. He lived in bunkers at Kapyong and Kumhwa. Miller was recommended for Non-Commissioned Officer (NCO) School and was transferred to C Company. When they found out he had Chemical, Biological, Radiological (CBR) training, they sent him back to B Company.

He went back on the line in the Iron Triangle. This was an important Communist Chinese and North Korean area and communications junction located between the cities of Chorwon and Kumhwa in the south and Pyonggang in the north. This triangle was critical for the Chinese entry into the south. It would be the scene of many battles during the war. Miller would learn famous names to come such as Pork chop Hill, Old Baldy, White Horse Mountain, T-Bone Hill and Arrowhead Ridge. While on Arrowhead Ridge, he could actually watch hand to hand combat happening on White Horse Mountain.

His company became attached to the 23rd Infantry Regiment along with N Company of a French Battalion. On October 6, 1952, they assisted the French in setting up a defensive perimeter after a Chinese officer had been captured with a map and orders of an impending attack on Arrowhead Ridge. At morning light, it came. Because they had completed the preparations in the evening, they were ready.

Mortars started pounding the encampment. A sergeant nearby was killed by shrapnel and a 2nd Lieutenant was hit in the elbow. His job was to make sure the bridges stayed open. Even with the shelling, the pontoon bridges stayed up. Part of the captured orders were that the Chinese were going to let the dam down the river open and flood the area below, destroying the bridges. Luckily it did not work.

His men had a Quad 50 halftrack vehicle or a vehicle with four fifty caliber machine guns mounted on it. Miller recalls the deafening noise of the battle as they fired constantly until the barrels actually glowed red. On the hill, the French held their ground, and the bridge was safe. As they gathered the next day to survey the carnage, they looked for people who may be alive. Surprisingly, a group of Americans had stayed in their bunker throughout the fight while the Chinese went back and forth. What amazed everyone including Miller was the number of bodies of dead Chinese soldiers that lay on the hillside. There were hundreds upon hundreds of dead. Arrowhead Ridge was the closest he had ever come to being killed.

On a road near Old Baldy, they were grading a damaged road when they heard booms behind them. Mortar rounds were falling nearby and had destroyed all the work that they had done. They came very close to killing Miller and his crew. In March 1953, they went back on the line relieving the 7th Division. Miller, being a short timer, was kept behind the lines working on roads. Eventually, he was sent home and discharged in May of 1953. He received numerous medal and ribbons for his outstanding service.

Miller arrived back home at Friendship Airport in Baltimore and was met by his Mom, Dad, and wife, Janet, whom he married January 27, 1951. He wasted no time in going right back to work with his dad, eventually taking the business over. He was also one of the first to be assigned to the licensing board for Carroll County. He is proud to say his number is #4.

Miller has one brother and one sister. He also has one son and two daughters along with five grandchildren. He retired from his business in 1991 and decided to finally retire with his wife of fifty-nine years. He maintains proud collections of memorabilia from his service including pictures, books, and Korean War icons. He has also maintained contact with a few of his comrades from the war.

Bill Hall

One thing he remembers about Korea is that it was cold, very cold!

Clarence Moore
Lieutenant J.G.
Aircraft Carrier CVL 24 Belleau Wood
U.S. Navy
World War II

Mr. Moore was born in the Howard Park section of Baltimore City on August 5, 1923 to Eugene and Lila Moore. He began his working career with the C & P Telephone Company in Baltimore eventually becoming a District Manager.

Clarence, like so many others, put his life on hold shortly after Pearl Harbor brought the United States into World War II. In February 1942, he enlisted in the U.S. Navy. Clarence received his boot camp training at the Naval Station Great Lakes. He eventually became a fighter pilot assigned to the Carrier CVL 24 Belleau Wood. Moore was assigned to Fighting Squadron 31 on the ship and flew the famous F6F Hellcat fighter plane. The Hellcat was a joy to fly. It could attain a speed of 376 MPH and had six 50 Caliber machine guns.

Serving in the Japanese theatre, their mission was to get ready for the coming invasion of Japan. During many missions their job was to protect the naval task force they were assigned to. He got to see a lot of action during his time in service. Their fighter group attached several facilities, factories, and airfields. He recalls also attacking the Japanese Battleship Ise, causing damage to the floating fortress.

One scary moment he had came when a Japanese Zero came down from behind him and fired, hitting his plane. Fortunately, because of the shielding behind the pilots seat, he was protected.

The closest he got to death was while on a mission protecting the task force. He was flying at twenty thousand feet when he was told to go to thirty thousand feet for a bogey. He chased the bogey as far as he could when his engine failed. The fleet radio operators got on the radio with him and dispatched a destroyer to head for the vicinity of Moore's aircraft. They told him to try to land in the water. He kept trying the engine switch and nothing happened. The ocean grew ever closer. Then he said God was watching over him. As he drove the craft right across the water, inches above, he tried once more in desperation to start the engine. It started and he was able to pull up safely.

Clarence Moore's greatest claim to fame happened on August 15, 1945. While on a mission, he fell in behind a Japanese dive bomber, shot two short bursts, and watched as the plane enveloped into flames and crashed into the ocean. It was the last day of the war. He is today in history books as shooting down the last enemy aircraft of World War II.

Clarence was discharged on December 31, 1945. For his service he received the Pacific Theatre Ribbon, Good Conduct Medal, National Defense Service Medal, and his third Air Medal with 2 Gold Stars. Moore achieved the rank of Lieutenant J.G.

Mr. Moore has one sister and two brothers. His older brother served in the U.S. Navy during the war while the younger served in the Merchant Marines.

After the service, he continued with C & P and lived in the Nob Hill section of Reisterstown, Maryland until 2006 when he moved into Carroll Lutheran Village in Westminster, Maryland. His wife of 63 years, Margaret Smith of Baltimore, died in 2008.

As of this writing, Clarence has departed this world. Today, he leaves behind three daughters, six grandchildren and two great grandchildren. When asked about what was his favorite part of being in the service, he said it was flying the Hellcat fighter plane. He loved it.

James L. Moore
Private First Class, Anti-Tank Company
349th Regiment, 88th Infantry Division (Blue Devils)
U.S. Army
World War II

Because the sergeant did not like digging and lying in foxholes, he would lay on top of the ground most of the time. His men wondered how he stayed alive. Suddenly a German shell landed nearby exploding in fury. The sergeant was hit just below the neck by hot shrapnel. PFC James Moore and a comrade crawled over to the injured soldier and began pulling him back to their position. As they did, the other soldier assisting Moore commented how bad the wound looked. The injured sergeant was taken away to be treated.

About a month later, another sergeant told the soldier that assisted Moore to go to the Command Post. When the soldier arrived, he found the formerly injured sergeant screaming in his face and lifting him by the collar yelling, "Don't you ever talk out loud about how bad an injury or wound is in front of an injured soldier."

Born in Smith County, Virginia on April 4, 1924 to James F. and Mary Moore, young Jim Moore worked the family farm. However, Jim was drafted into the army on February 8, 1943. He was shipped to Camp Howes, Texas where they were put in temporary tar paper barracks leased to the government by film and music star Gene Autry. He was assigned to a rifle company in the 86th infantry.

Following boot camp, he was sent to Fort Sam Houston in San Antonio, Texas and assigned to an anti-tank company in the 88th Infantry Division. The 88th would be known as the first all-draftee division in the army. Then it was off to Newport News, Virginia where they were loaded up on the Empress of Scotland, a formerly converted cruise liner, and on December 17, 1943, they sailed off for Africa.

They arrived in Casablanca on Christmas Day, 1943, did mountain training for a few months, and went on to Oran in Algeria. Naples, Italy was next. They spent a month there waiting for enough equipment for a convoy. In April 1944, they were moved to the front lines. The men of the 88th would see some of the hardest-fought battles of World War II.

On June 4, the 88th was the first division to enter the "Eternal City:" Rome. Moore believes that this was the first time in history that Rome was captured from the south. When asked what major battles he fought, Moore humbly tells that all of Italy was a major battle. The Germans always had the high ground because of the mountains. It was tooth and nail all the way up through Italy. They would be on the line for 15-30 days at a time and then were allowed to go back and rest for a week or two.

There were times when the soldiers had to resort to other means to get supplies moving. Such were the times when they used mule-trains to move up the mountains due to the harsh terrain and mountains. At one point, Moore was the radio operator. An officer called them on the radio asking where they were. Moore replied, "I am pulling a mule and holding the tail of another. I am cold, and it is raining cats and dogs. How am I supposed to know where I am?" The officer called back wanting to know who was on the radio. Obviously, no one answered.

Jim More related that he lost a lot of good buddies. One thing he related several times was that they all depended on each other. His biggest inner fear was he never knew when it was going to be his turn. One of his best friends, a lieutenant, got shot in the head on the last day of the war.

Once the war was basically over, they were in the Po Valley cutting off the Germans trying to go to Austria. German soldiers heading back to Germany were lining the roads still carrying their weapons. The Americans would drive right past them but were told if anyone shot at you, shoot to kill. They stayed at a resort in the Brenner Pass until the rear echelon took over the hotel. It was odd to see American and German soldiers guarding the doors.

In the Alps, civilians would point out to the Americans where German soldiers were hiding out. In the evenings, the Americans would sneak out and capture the enemy holdouts. One memory etched solidly in Moore's mind was in the northern part of Italy. As they would eat their rations, old men and women and young children would watch them eat. The civilians held containers in their hands hoping for food from the soldiers. Moore stated that many times they would go hungry because they gave their food away.

During his time, Moore got in trouble, mostly for driving women around with his lieutenant on their jeep. But his reputation with the motor pool paid off as he was pulled out of trouble and put back in good graces. The young soldier was part of a special detail for two months guarding 100 German soldiers re-building a railroad. During their time, the 88th would see such battles as the Gustav Line, Mt. Damiano, Volterra, the Gothic Line, Belvedere, Mt. Grande, Monterumici, Verona, and Vicenza.

At last, with his 85 points, he was sent home and eventually discharged October 27, 1945 from Fort Jackson, North Carolina. For his service, he received the European-African-Middle Eastern Ribbon, three Bronze Battle Stars, Combat Infantry Badge, and the Good Conduct Medal.

After the war, he worked for an electrical contractor until he became Inspector for Montgomery County, Maryland for thirty-one and a half years, finally starting his own business. Today he lives in Mt. Airy, Maryland with his wife Judy of forty years, three children, six grandchildren and six great grandchildren. He is also an active life member of American Legion Post 171 and the VFW in Mt. Airy, Maryland. When he returned home, he knew two things: he did not want to farm or fight again.

Robert Murray, Corporal
Tank Scout, Machine Gunner
3rd Squad, 4th U.S. Cavalry,
25th Infantry Division
McKenzie's Raiders,
United States Army
Vietnam War

After only one week in Vietnam, PFC Robert Murray met one of the most formidable enemies he had ever met...one that took him down! While on foot and checking for land mines, he was attacked, hurled in the air, and landed hard on the ground by a water buffalo. Shortly after regaining the upper hand, he shot and killed it.

Born in Wilmington, Delaware to William and Florence Murray on December 14, 1945, Bob was a normal kid advancing through high school. In October 1965, he enlisted in the U.S. Army. He went through basic and advanced training and within two weeks from graduation, he was in combat. The young corporal was assigned to the 3rd squad, 4th Cavalry, 25th Infantry Division as a tank scout and machine gunner. The men were known as McKenzie's Raiders.

Murray was assigned to Cu Chi, north of Saigon and was part of the largest army division in Vietnam. Their assignment was between Cu Chi and Tay Ninh. Their missions would be between two to three weeks on an armored personnel carrier (APC) and they would scout ahead of the infantry and usually saw combat. Many times, they would be assigned eight-man squads doing ambush patrol. This was the scariest. A squad would leave at dusk, set a perimeter, and report movement by going out only one to two miles. Very seldom did they sleep on cots. He remembered the words of his father telling him to join the Navy or Coast Guard because you would always have a bed and a hot meal.

Once Bob tells of a patrol where they had a wood line behind them and fifty yards clear out in front and were dug in with two machine guns. Around 2 AM, they heard noise behind them. They saw through a starlight scope a company of Viet Cong, about 100 men, coming out of the woods behind them. Luckily, they stayed quiet, and the enemy passed.

Murray talks about the politics of the time. After coming back from a mission on Christmas day at 10 AM, the squad was hit by sniper fire. They returned fire, Murray with his 50 cal. The officers chewed them out and wanted to know why they violated the cease fire for Christmas. Each man in the squad had to make a statement.

One of his memorable moments was attending the Bob Hope Christmas Show. But what was more memorable later was that in the book, *Tunnels of Cu Chi* by Tom Mangold and John Penycate, the Viet Cong were right under the show listening to the whole thing in a tunnel.

Bob had two R & R, Rest and Relaxation, passes while in Vietnam. One was to the town of Vung Tau and the other to Japan. While in Vung Tau, he found out that the day after he left, his squad had been wiped out on a mission with only one surviving.

During another mission in March 1967, another APC was following their tracks, hit a land mine, and flipped over. Twenty-one-year-old Donald White was trapped inside and burned to death. Sgt. Walter Blalock, machine gunner, died after flying out. He pulled medic Dennis Jensen and flew him out also. All three soldiers died. All of this was being done with munitions exploding and sniper fire all around him. Murray was put in for the Bronze Star but never received it.

The most horrible incident he remembers was one evening they were hit by a regiment of North Vietnamese Army (NVA) while on bunker line patrol around Cu Chi base camp with a company of the 14th Infantry. In the morning, they had to go out and retrieve the 14th, 100 men, all killed in action. After loading the bodies onto five-ton trucks, he watched them pull away as blood gushed out the backs of the vehicles. He left Vietnam in April 1967 and did NCO training at Fort Dix, New Jersey for the remainder of his career. He was discharged in October 1968.

Vietnam Vets never got the respect they deserved for their service. Finally, today things seem to have changed. He started feeling better about his service after Desert Storm. He was never so proud when he attended a Grand Ole Opry to hear Jan Howard sing. During the performance, she acknowledged Bob and war buddy, Daniel Carlisle before the entire crowd. She, too, had lost a son in Vietnam. They stood and cheered for them. A more recent honor was August 2009 when the Baltimore Ravens football team, at training camp, acknowledged about 500 uniformed veterans their service.

Of all the medals Bob received, his proudest is the Combat Infantry Badge. This is his badge of honor. He did receive two minor wounds but received no Purple Heart. He recalls his most memorable moment was getting his orders to return home and having survived.

Bob moved to Carroll County in 1970 and sells and promotes erosion control products. He lives in Westminster with his wife Lynn, of 31 years and their five children. He has one brother, William. Bob bears many good and bad memories of the unpopular war but was willing to step up and serve his country even when it wasn't the accepted thing to do. This is another Vet who certainly did his share.

William E. Myers, Sergeant
8th Air Force, Second Air Division
448th Bomb Group
World War II

William (Bill) Myers was born November 14, 1921 in New Windsor, Maryland. He worked several jobs during and after school including being an ambulance driver for Hartzler Funeral Home, working in a slaughterhouse, and a canning factory. He eventually went on to Western Maryland College where he spent three years majoring in accounting and also taking ROTC training.

When World War II broke out, Bill enlisted in the Army Air corps in 1942 at the age of 21. He bypassed the infantry and attended the Aviation Cadet Program. He attended training in Boise, Idaho. Unable to pilot due to a physical problem, he learned gunnery and was assigned as a tail gunner to a B-24 Liberator Bomber and stationed in Seething, England with the 448th Bomb Group. He arrived on a troop ship from Boston. His 10-man crew would have to fly 25 missions to earn the right to go stateside. In the words of Captain Robert Morgan of the famous *Memphis Belle*, "Twenty-five doesn't seem like much until you start flying them." Bill agrees. He also stated that the movie was very realistic.

In his book *Achtung! Noon Balloon*, Bill relates that "after several training missions, delay from bad weather and more snow which was removed from the runways, the crew finally flew its first combat mission on January 13, 1945. It was in support of the Battle of the Bulge in the Ardennes."

On their flights, Bill would go to the tail gunner position with his fifty cals. He wore heavy boots, a heated suit under his flying suit, and a handy parachute nearby. He wrote in his book that "the crew named their plane the *Achtung! Noon Balloon*." The name was chosen because most of the bomb runs were over the target area around noon time. The aircraft was commanded by Major Kay Flinders. Each crew member had their nicknames also; names such as Skipper, Mac, Buck, Flaps, Mo, Mitch and Hank. Bill's, because of his stature, was Pee-Wee."

At Seething, there were three flight crews per Quonset hut. Bill remembers how they knew everyone in their huts and on one day, one of the crews never returned. They were killed in action. On a mission over Dunneberg, Germany the crew was bombing a dynamite factory. After the payload was released, a sharp concussion slammed the Liberator making everyone think they were hit but it was the actual shock wave from the explosion below them.

Bill wrote in his book that "no one can recall the exact date, but all of the crew remembers the day of the "near miss," when bombs from a plane just above them in the "high element" came very close. The plane dropping the bombs was camera-equipped and recorded the near-miss on film."

As a tail gunner, Bill got to see a first in aviation. As he manned his guns during a mission, a new aircraft zoomed in for the kill. It was a German jet. No one had ever seen one before. He felt he got some good shots in but did not see a shoot-down. He related that this one really scared him. Later, on the ground, they were warned that the Germans might have a jet aircraft in the skies very soon. On April 25, 1945 the *Achtung! Noon Balloon* flew its last combat mission in the 448th Bomb Heavy Group.

When V-E day came, his crew began training for the war in the Pacific, but the war ended before they would fight again. The only crew member lost was the flight engineer who was on a training mission with another flight crew.

After the war, Bill graduated from the Baltimore College of Commerce with a degree in accounting. He met and married Irene Van Fossen. They had met back in college and through 448th reunions. Bill was the Comptroller for the Community Chest of Baltimore, the Baltimore City Public Housing Agency, and Auditor for the United Fund. He also worked for Baltimore City being in charge of the Retirement System. They purchased a home in Owings Mills, Maryland and Bill became active in their fire department where he served as secretary and treasurer. He also helped organize the New Windsor Volunteer Fire Department and served there also as secretary and treasurer. He began and maintained for years his own tax and accounting service.

He is the proud father of two daughters, Pamela and Patricia, and one son, William Douglas. He also has six grandchildren. Bill Myers is retired from his accounting and tax service. He lives at Carroll Lutheran Village in Westminster, Maryland. His time in the military as a part of the Greatest Generation is quite memorable and his service to his country at a terrible time in history is truly appreciated.

Michael Peters, Sergeant
U.S. Marine Corps
2nd. Battalion, 4th Marines,
3rd Marine Division
"Magnificent Bastards of Vietnam"
Vietnam

The fourteen-man squad headed into the jungle to set up a listening post near Vin Quan Thong. As they did, Michael Peters had them set up for covering fire. The two men setting up the post to monitor the North Vietnamese Army (NVA) went forward. The squad saw movement and Peters called the men to run back and they would fire over their heads. Once they began to shoot, they were peppered with enemy fire. They had walked into a classic NVA "V" ambush.

With his father having been in the Marines, Michael Peters had only one dream…become a Marine. Born to Allen and Helen Peters in June 1948 he graduated in June, 1967 from Glenelg High School in Howard County, Maryland. The want-to-be marine was told by his dad that he wasn't good enough to be a marine. But he knew what his dad was doing. Peters went for his physical at Fort Holabird in Baltimore and was told by a doctor that he was partially blind in one eye and could not be accepted. Peters pleaded with the doctor to approve him because he wanted to prove to his dad that he was good enough. The doctor relented and approved him for service.

After 30 days in Camp LeJeune, NC and 30 more days in Camp Pendleton, CA, the young soldier was ready. He remembered getting handed coffee and cookies by the U.S.O. ladies before he left. One touching moment for him was, as they were leaving San Francisco, he looked back at the night lights of the city from the aircraft and wondered if he would ever see them again.

He arrived in Da Nang, Vietnam in December 1967. He noticed that there were some soldiers wearing their helmets with the words "Two days and out" written on them. He found out that they were leaving Vietnam in two days so he wrote on his helmet "395 days and out."

He and his men were assigned to the helicopter carriers U.S.S. Iwo Jima (LPH 2) and later the U.S.S. Princeton (LPH 5), to move up and down the coast around Quang Tri delivering amphibious assault Marines where they were needed. Peters and the Marines were sent to the Philippines while on the Iwo Jima for 30 days of jungle warfare training. It would later prove to be valuable.

Peters tells the story of a Viet Cong (VC) sniper in hidden in a church steeple raining fire down on him and his squad who were hiding behind a mound. They were taught never to shoot over a mound because that is likely where the enemy will fire. They were told to shoot from the sides of the mound. He remembers an RPG (rocket propelled grenade) hitting the top of the mound and looking up and seeing dirt and debris flying over them. They had access to an ONTOC vehicle. It is a 106 mm self-propelled rifle M50 anti-tank vehicle nicknamed (The Thing). It is mounted with six 106mm recoilless rifles as its main armament and could fire in rapid succession. One shot took out the steeple ending their dilemma.

Nicknames were mostly used in Vietnam because men did not like to get to know their buddies too personally. They never knew if they would make it through the war. Peters had friends named Townson and Sugar Bear. It was a way of trying to help keep your sanity.

During the Tet Offensive of 1968, the squad was to set up a listening post near Vin Quan Thong to monitor communications with the NVA. The fourteen-man squad would cover the two-man team. Peters carried a radio.

Movement ahead caught their attention. Peters called the monitoring team to come back and they would begin firing over their heads to cover them.

As soon as they opened up, they received fire from all sides. They had walked into an ambush. The NVA had used their classic (V) to all but surround the Americans. The fire was intense. Peters' friend, Corporal Townson was shot in the head. Peters dragged the wounded soldier to cover while he himself was shot in the arm and leg. He fell and realized two rounds went in the radio. He fired a phosphorous round into the radio to totally destroy it and prevent it from getting into enemy hands.

Sugar Bear had been captured in the fire fight. They could not get to him to rescue the wounded Marine. The fight subsided and night fell. The battle-weary Marines waited through the night for help. They could hear the cries and screams of Sugar Bear as he was being tortured by the NVA throughout the evening. Helplessness fell upon them.

The next morning, the NVA were gone. They found Sugar Bear. He was dead: his body mutilated. They found out later that the NVA was building for a massive offensive for April. Peters was discharged in May 1970 after being given choices of Drill Instructor School, Embassy Duty, or an early out. He chose the latter.

Coming home was not pleasant. No one seemed to want to hire a Vietnam Vet. He was told things like he wasn't stable, he had committed war crimes, and they didn't know about how he would handle his temper. He, did, however find a friend at the W.R. Grace Company where he has worked since 1970. They also told them that they were proud of him and what he did.

Peters received 2 Purple Hearts, Bronze Star, Bronze Star with a "V," Cross of Gallantry, Presidential Unit Citation, Navy Citation, Good Conduct Medal, National Defense Service Medal, Vietnamese Service and Vietnamese

Campaign medals. His proudest moment was when the Commandant of the Marine Corps awarded him his medals.

Today the proud Marine lives in Westminster, Maryland with his wife Lura, of 25 years, two daughters and a son, six grandchildren, five grandchildren, and seven grand dogs. The Marine says that he is proud of what he did in Vietnam no matter what anyone else thinks. He loved the challenges of battle.

Robert Peterson
Hospital Corpsman, Technical Sergeant
United States Army
World War II

Robert Peterson was born on October 6, 1923 in Frederick, South Dakota, the son of Edward C. and Elizabeth Peterson. There he was raised most of his young life until he enrolled in Northern State Technical College located in Aberdeen, South Dakota.

After World War II broke out, he found himself signing up for the draft. On March 10, 1943, the twenty-one-year-old enlisted in the U.S. Army and received boot camp at Camp Barkley, Texas. The Army opened specialized training, so Bob took the engineering course. They sent him to Western Maryland College in Westminster, Maryland with 400 other troops in June 1943 to accomplish this. He had also applied for the Air Corps and made Cadet, but they cancelled the course. He opted for Medical Basic. Bob went to school as a surgical technician/hospital corpsman.

Upon graduating, he was assigned to the 74th Field Hospital Unit (similar to a MASH unit). The unit had 182 enlisted and 21 officers. He was sent to Hawaii for a six-week course in anesthesia. Sixteen days after the invasion of Okinawa in April 1945, Bob Peterson arrived. He commented that in the 83 days of battle, they received 1,844 patients. He remained on the island for six months after the war was over. Bob is amused that after only six weeks of training with anesthesia that he was giving it to patients if the doctors were too busy.

The medic was recalled in the inactive reserves during the Korean War but remained for only ten months. After that, he was assigned to Fort Meade in Maryland for five months teaching basic medical. He was discharged on March 15, 1945. He states it was "three years, five days, and one-half hour."

He recalls his first case. It was a Finnish boy from Wisconsin who was brought in with his chin blown off. He could not breathe due to his injury. The doctor told the personnel to flip him on his back. After that, a tracheotomy was put in his throat and the young soldier was able to finally breathe, eventually being evacuated to Guam. He also remembers a soldier shot by a sniper. He had a lateral wound through both sides of his abdomen. He clearly recalls the doctors patching him up.

Probably the most notable one Bob can recall was a 2nd Lieutenant from California who had been in a foxhole. A hand grenade had been tossed in and he was unable to get out fast enough. The explosion took one leg off at the knee, the other above the knee, and both hands. Bob was responsible for going through his effects and found a picture of the young man's wife and child. The young man survived. He says that this one bothered him badly.

For his service, Bob Peterson was awarded the Asiatic Pacific Campaign Medal, the Good Conduct Medal, and the National Defense Service Medal.

Peterson was not the only member of his family to fight for his country. Of his five brothers, the oldest was killed in action over Germany. He was a B-24 Liberator Pilot. The youngest brother served in Korea as an F84 Fighter Pilot.

One of his many talents which he shared was his love of woodcarving. His home is filled with and numerous articles have been done about his talent for carving natures creatures. He has several in his home and many more out there in the unknown.

In April 1944, he met his wife, Pauline Dunn of Westminster, Maryland, and was married for 63 years. They have one daughter and two sons. After the death of his wife, he met and married Sylvia Creamer. They were married in

May 2009. Bob worked for many years at Tevis Oil in Westminster and is now retired and living in Carroll Lutheran Village in Westminster with his new bride.

Many soldiers were undoubtedly saved and rescued from the jaws of death by this humble man. The skills he learned in the Army were very much respected. It is good to see someone who fought in the greatest war ever, live a long, healthy, and happy life. If you ever see Mr. Peterson around, tip your hat to him and say "Thanks!"

John D. Rill
Sergeant
Co. H, 2nd Battalion,
120th Infantry Regiment
30th Infantry Division
U.S. Army
World War II

The freezing cold of that winter of the Battle of the Bulge kept the troops just trying to stay warm. The Germans were making life miserable for the Americans. Battles erupted all over. PFC
John Rill, ammunition bearer in the 120th Infantry Regiment huddled down during a fire fight. Suddenly mortar rounds found their marks as they fell into the encampment. The first one killed a sergeant. The next one took out the First Gunner followed by one that just missed Rill but killed the other ammunition bearer. In all they lost a whole section including the Lieutenant and runner of the platoon. That night a sergeant took over as Lieutenant and Rill was made Corporal and would be the runner for his outfit. The next day, the replacements finally came in.

Born on March 29, 1917 to Albert and Laura Rill of Hampstead, Maryland, young John Rill had three brothers and six sisters. The Manchester High School student graduated in 1935 in a class with 13 girls and 3 boys. He worked for National Advertising painting signs. The twenty-six-year-old was then drafted in the fall of 1943 into the Army.

Rill spent his early days in boot camp at Fort McClellan, Alabama and was shipped immediately overseas to England on the Queen Mary converted cruise liner. He would be assigned to H Company, 2nd Battalion, 120th Infantry Regiment of the 30th Infantry Division.

After a large section of the 30th landed on D-Day, the remainder of the division arrived five days later. Rill recalls climbing down the side of the ship onto the landing craft and rushing into shore. Normandy's beaches still showed heavily the damage the American forces had taken upon the initial landing. But for Rill and the other troops, this was only the beginning of some of the most ferocious fighting of the war.

The Germans wasted no time in letting the Americans know they were not happy with them being there. He and the American soldiers were thrust right into battle. The G.I.'s fought inch by inch in the thick hedgerows of Normandy making their way slowly south into France, fighting town by town all the way. Rill's regiment stayed north of Paris and headed east aiming for the Siegfried Line.

The 120th became part of Task Force Harrison. The route of the Task Force was from Pontoise northeast of Roye, Peronne, Cambrai and Valenciennes, crossing the Belgian border and became the first allied troops to enter Belgium, capturing the town of Tournai.

Rill would make his mark during the seizure of Kalrath, Germany in February 1945. Company H machine gunners pushed forward to place their automatic fire protection at the ends of the town. Corporal John Rill and his sergeant found a perfect spot for their machine gun to cover a road leading from town. Still the pair received fire from enemy troops. On their own, they decided to clean out the houses in the way. Leaving their crew at their machine gun, the two G.I.'s went through the houses and before they were finished, captured thirteen German soldiers. For their heroism, they received the Silver Star.

They then headed north of Cologne to the Rhine River. They crossed on outboards while the Corps Artillery pounded the opposite shore as they crossed. The Americans moved very quickly on shore; so quickly that they ran past

German soldiers on the way. The Germans were picked up later surrendering to the Americans.

For weeks the 120th would battle their way to the Elbe River. During a battle, his Section Sergeant was hit by fire from a German aircraft strafing the troops. Rill was made Sergeant. He was called upon one time to run wire from the gun emplacements to Headquarters so they could have radio communication.

Once they reached the Elbe River, they waited for the war to end. Everyone knew it was almost over. They also waited for the Russians coming in the opposite direction. Even though they were allies, they were told not to associate with the Russians. Finally, the war was over.

One of the first things many of the G.I.'s wanted to do was to go to Paris. Many of his buddies got passes to go but he would not because he had a wife and kids at home. John Rill came back to the States on the Queen Mary, the same ship that took him to war. He arrived at Fort Meade, Maryland and called his wife to come and get him. He was discharged in September 1945. He received the Combat Infantry Badge, Good Conduct Medal, 3 Battle Stars, a Silver Star and several other medals and ribbons for his service.

After the war, Rill went back to National Advertising. The company was bought out by 3M. He stayed and thirty five years later retired. He has also been a member of the Arcadia Volunteer Fire Department, entering their Hall of Fame and has received the Volunteer Citizen of the Year Award from them. Rill is also been a Member of the American Legion Post 200 in Hampstead, Maryland for fifty years.

Retired today at the ripe old age of 93, he enjoys gardening and raising vegetables and flowers at his home in Finksburg, Maryland. He has three boys and two girls, twelve grandchildren, fourteen great grandchildren, and seven great, great grandchildren. He was married to his wife

Ella, whom he met at a dance, for fifty-four years until her recent death. It was very apparent during this interview that his family is very proud of him and he, too, is proud of what he did in his service to his country.

Marshall Rose
Sergeant, Army Air Corps
Waist Gunner, B-17
381st Bombing Group,
535th Bombing Squadron
8th Air Force
World War II

On May 24, 1944, the "Return Ticket" B-17 Bomber was taking Staff Sergeant Marshall Rose and the nine-man crew on their first big bombing mission. Their target: Berlin. At 25,000 feet and -30 degrees, they dropped their bombs and turned to head back. Out of nowhere 60-75 enemy aircraft came. A bomber on the port side was hit and crashed into another bursting into balls of flame. A German fighter crashed into the wing of the "Return Ticket" fatally damaging the plane.

Rose was thrown onto the floor with the ammunition rack on top of him. Another death throe and it rolled off. With the plane spinning, he was able to get his parachute on, got to the exit door and kicked it open. As he fought to get himself out the door the chute accidentally opened causing the chute to go out the door and snag onto the tail. He was yanked out the door with his body weight as it finally pulled loose.

As he floated to the earth, he noticed the plowed field below him just a few miles from Berlin. Upon landing, two farmers with pitchforks chased him in the 15 miles per hour wind across the field. After he stopped being pulled by the chute, he looked up at a German pointing a rifle at his head.

Born in the Washington, D.C. area to Joseph and Isadore Rose on July 12, 1923, Rose would eventually graduate from

Roosevelt High School in 1942. He entered a civilian aircraft school and was eventually drafted in March 1943.

The nineteen-year-old received basic training in Atlantic City, New Jersey followed by aircraft mechanic school in Newark. Upon finishing, he went to aircraft gunnery school at Tyndall Field, Florida and then to Drew Field, Florida where he was assigned to a crew to take B-17 training as Assistant Engineer-Waist Gunner.

The newly assembled crew went to Savannah, Georgia to pick up a new B-17 and take it overseas via Goose Bay, Labrador. Upon arriving in England, they had to give up the new B-17 and get an old, beat up one. Then assignment came to the381st Bomb Group, 535th Bomb Squadron, 8th Air Force.

Their first mission was going well. The "Return Ticket" saw Berlin come into view and they dropped their bombs on their targets. The P-51's escorting them had no problems with enemy planes. Then, as they began to turn to go home, out of the sun came 60 to 75 German fighters. Before the P-51's could fight them off, the Germans had damaged several ships. Two bombers collided and blew up. Another four were lost in action.

After Rose was taken prisoner, he was sent to a local jail and then to a large barn that housed about 30-40 airmen. It was there that he had seen three of his crew. He knew then in his gut that the other five had gone down with the plane.

The POWs were then taken by train to Berlin where they boarded a train. He was surprised to see the German soldiers holding off the civilians to keep them safe. There was about twenty POWs in his group and had about eight guards with Burp guns. They boarded the train and moved to an interrogation center. After giving only the allowed information, he could hear one of the guards saying," I see they're scraping the bottom of the barrel and sending boys." He would become POW #1631.

They were sent to Stalag Luft IV, a new camp along the Baltic Sea in Northern Germany, arriving three days before the arrival of food. Life at the POW camp was unbearable. Lice and fleas were rampant. Many became infected with pneumonia, diphtheria, pellagra, typhus, trench foot and tuberculosis. The buildings had twice as many men as they should have. Many slept on floors or stools pulled together. They would go months without showers. Rose remembers one POW being shot while trying to climb the fence to escape.

On February 6, 1945, they were told they were being evacuated for a three-day trip. They could see the flashes of Russian guns from the east. It wound up being three months for most. They walked 500-600 miles in freezing cold, sleeping in barns or in the open. They drank water from ditches or ate snow when available. Diarrhea and dysentery were the main problem. The average man lost 30% of his body weight. When he was freed, he weighed less than 100 pounds.

Finally, in the spring of 1945, it was over. British troops surrounded them and the Germans. They walked past mine fields and blown-up bridges. Rose recalls carefully crossing a blown-up bridge on its trestles. The POWs were sent to France for a month to fatten them up before returning home. He and all of the POW's received numerous shots before being released. On November 7, 1945, Marshall Rose was discharged from the Army Air Corps receiving four European Theatre Operations ribbons and the Good Conduct Medal.

Returning home, he was treated as a hero. However, it did have its prices. He broke up with his girlfriend. He said she was still a teenager, but he was now grown up. But he did marry his 6th grade sweetheart, Lillian Rose, his wife of 49 years. She was an Army nurse in the Pacific. She died in 1999. He worked for the National Bureau of Standards for ten years and went on to Johns Hopkins Applied Physic Lab until retirement in 1992.He recently had a Commitment Ceremony with Leslie Crow. Rose moved to Carroll County in 2007 and has made it his new home. He has four children: three boys and a girl. Also, he has eleven grandchildren and four great grandchildren.

David Schaeffer
Lt. Colonel, U.S. Army
Tenth Army Headquarters
Pacific Theatre
World War II

After the Nazi invasion of Poland September 1, 1939, young David Schaeffer told a friend that they were soon going to be in a war. His prophetic words became reality for him in June of 1941. He was born to Noah and Elizabeth Schaeffer of Westminster, Maryland June 9, 1920. David, who had been attending Gettysburg College, entered the ROTC program, and in June 1941, at his college graduation ceremonies, was commissioned a 2nd Lieutenant in the Army, receiving immediate active-duty orders to Camp Lee, Virginia.

In January 1942, he was transferred to the 82nd Infantry Division and served as a platoon leader of the Quartermaster Truck Company. In late 1942, he was transferred in the cadre to activate the 98th Infantry Division, receiving a promotion to Captain and command of the quartermaster company. In early 1943, he again was part of the cadre to activate the 63rd Infantry Division which trained in Florida before moving to Camp Van Dorn, Mississippi in September 1943. He was again promoted to Major with assignment of Assistant Division Quartermaster.

In June 1944, he was ordered to Fort Sam Houston with assignment as a staff officer of the newly activated Tenth Army Headquarters, destined to be in command of the Expeditionary Task Force to invade and seize the island of Okinawa from the Japanese forces.

Before leaving for the Pacific, he married Ms. Bobbye McDaniel on July 1, 1944 in Baton Rouge, Louisiana, having met her at an officer's club dance at Camp Van Dorn in October, 1943.

The Battle for Okinawa began April 1, 1945 and required 82 days before the island was secured, resulting in the destruction of many thousands of Japanese, fighting to the last man. Schaeffer received a promotion to Lt. Colonel as preparations and planning for the invasion of Japan's home islands were already in progress. The dropping of the Atomic Bombs, however, resulted in the surrender of the Japanese, and formal ceremony and signing September 7 to General MacArthur.

The Tenth Army Commander received the surrender of the commanders of the Japanese forces on the other Ryukyu Islands, and Schaeffer was a member of the task force sent to supervise the inventory of troops, supplies, and munitions which had to be destroyed.

For his service, Schaeffer received the American Defense medal, the WW II Victory medal, the American Campaign medal, the Asiatic-Pacific Campaign medal, and the Army of Occupation-Japan medal. He was discharged in March 1946 relieved of active duty but remained on inactive duty with the Army Reserves until the end of the Korean War in 1953.

David was not the only one in his family to be a veteran. His younger and only brother, Edgar, served in the European conflict as an officer in an armored division. Edgar had also served a tour of duty in the Korean War. Schaeffer's son, David, served as a 1st Lt. in the Army in the Vietnam War.

He returned home in March 1946 and started the Schaeffer Lumber Company in Westminster with his Father and Brother, retiring in 1986. Upon retirement in 1986, he moved to Lake Wales, Florida, while spending summers in Westminster visiting with family and friends.

They have one daughter and two sons. The children have given them four grandsons (two deceased) and they have three great grandchildren.

Bill Hall

For a man who has given five years of his life in the defense of his nation, and several more on inactive duty, one can never say "thank you" enough. It is people like these who have given us the lives and protected freedoms we enjoy today. God bless them all.

James Schumacher, Jr.
Staff Sergeant
B Troop, 1/158 Cavalry
58th Infantry Brigade Combat Team
Operation Iraqi Freedom

Staff Sergeant James Schumacher, Jr. arrived at Camp Cropper in Baghdad in 2007. Part of the job his unit would play was critical site security and detainee operations. Their job was to keep the prisoners of the war in and the terrorists out. The soldier's main protection from the outside of the camp was a sniper screen. Keeping the Sunni and Shia prisoners away from each other was the hard part. It bothered him most knowing that when he looked into the prison population, he would be looking face to face at murderers behind the fence.

Schumacher was born May 10, 1971 to James and Karen Schumacher of Carroll County, Maryland. After graduating from Francis Scott Key High School in 1990, he went to work for the Town of Sykesville. Wanting to move forward in his life, he enlisted in the U.S. Army in 1997.

The young recruit took basic training at Fort Jackson, South Carolina. He moved to Aberdeen Proving Ground for Truck Vehicle Repair School and gained a job skill. After that he went on to work in a battalion staff job as an S2-3 Clerk training in security.

In 1998, he was dispatched to Korea with the 2nd Infantry Division, 226 Maintenance Battalion, working on all equipment and vehicles including wheeled and tracked. The 16-hour trip to Korea was a long flight. When the plane started down, he saw Japan and Mt. Fuji. The exotic country of Korea let him know he was not in Kansas any more. After arriving in Seoul, he remembers jumping in a cab with some other soldiers and looking at all the lights and signs and nothing being in English. He recalls riding past the Olympic stadium and being impressed by its size.

While in Korea he, on a daily basis, led soldiers in force protection exercises designed to defend against an attack by more than one million soldiers of the Korean People's Army stationed on the north side of the Korean Demilitarized Zone. He trained and led Korean augmenter's (KATUSA's) and helped protect the Republic of Korea Army as a member of the forward deployed 2nd Infantry Division.

After deployment to Korea, he went to Fort Sill, Oklahoma in 1999 until 2000. He was part of the 3 Corps Artillery. They worked on weapons such as the Paladin, howitzer and the Multi-Launch Rocket System (MLRS). Schumacher would then be discharged in late 2000.

He went into the National Guard in 2001 assigned to 13 Bravo as a cannon crew member which was part of the Alpha Battery, 2nd/110th Field Artillery. In 2002 until 2003 he participated in Operation Noble Eagle. That was the Homeland Defense Force Protection in the military district of Washington D.C. It would involve Fort Myers and Fort McNair in D.C. Schumacher served as a Military Policeman (MP) and was on several missions. Two missions he is proud of were doing security for President Bush's Fitness Challenge where he got to meet the President. The other was serving during the inauguration of President Barack Obama.

In 2003, he went to M-Day status where he would work the 1 weekend a month and 2 weeks a year. He worked for the Maryland Military Department doing Anti-Terrorism Force Protection and security at Camp Fretterd in Reisterstown, Maryland for 4 years.

Duty called him up to Operation Iraqi Freedom from 2007 to 2008. He served at Camp Cropper in Baghdad doing critical site security and detainee operations. Iraq revealed a culture shock for Schumacher. He was amazed by the Biblical landscape and buildings a thousand years old. He saw castles that Saddam Hussein had built and plush palaces that were all but destroyed.

While in Iraq he gained firsthand tactical knowledge of intelligence operations, terrorist cell and network operations, geographic information systems, and information assurance. One thing he is proud of is that his entire troop came back home. Schumacher recalls the camaraderie expressed by the National Guardsmen. They were all trained three months before leaving for Iraq in convoy security, weapons training, and urbanized terrain training.

In 2008 he returned to Camp Fretterd. In February 2009 he worked for the Education Benefits Department at the Fifth Regiment Armory in Baltimore helping GI's get help applying for the benefits they so richly deserved. In May 2009 he was assigned to the General Henry Evans National Guard Armory in Westminster, Maryland as a recruiter where he still serves today.

As part of his job being a recruiter, he cannot understand why there are not fifty people lined up every day at their door. Schumacher believes that the military is a great option for those who do not know what they want to do or are having problems getting a job. By joining the reserves, a person gives up one weekend a month and two weeks a year for benefits such as the G.I. Bill, schooling, skills, training, and a can-do attitude. He says that the military is "not a hand-out but a hand-up." He also states that for himself, the first enlistment is you owing the Army but after the first one, you are part of the organization and what makes it work. It is one of those organizations that can really help to put a person on track.

Schumacher has lived in Carroll County all his life and currently resides in New Windsor with his wife, Jiyeoun, of 9 years. They have no children. However, Schumacher has 2 brothers and 3 sisters. The Army was there at the right time in his life and helped him to become the man he is today: a proud warrior who loves serving his country.

Richard Sheedy
Sergeant
Communications Radio Mechanic
Army Air Corps
780th Army Air Force
World War II

Young Richard Sheedy, a radio mechanic in the Army, knew someone had to go up to the top of the large "homing" tower to replace the burned-out light. Someone volunteered and went up. The transmitter was turned off. Sheedy was in the shed below and began disconnecting the wires but neglected to make sure that the current was off to the #8 wire. As he grabbed it, electricity raced through his body. The current coursed through him until he was able to lift his feet from the ground. His weight released him and he fell to the ground.

Richard Sheedy was born July 10, 1919 in Dunkirk, New York to Frederick and Lillian Walters Sheedy. He was raised there with his three brothers. For two years he worked as an apprentice tool maker for the American Locomotive Company. While there he learned to read blueprints and learned a technical background.

He recalls while in Dunkirk, standing on a corner December 7, 1941 at Central Avenue and 4th Street in front of a hotel, a friend approached him and declared that the Japanese had bombed Pearl Harbor. In March 1942, he took the exam for flight school with the Army Air Force.

Before leaving Dunkirk, his mother gave him a letter from her nephew given him by a chaplain. It mentioned his brother Robert who had been wounded in Europe. The chaplain stated that his brother Bob was the bravest soldier he had ever seen. Sheedy thinks very highly of his brother and considers him a true hero. The sailor thinks to this day how that must have affected his mother having two sons fighting in the war.

On June 9, 1942 he took the oath of office as an aviation cadet in Buffalo, New York. Sheedy wanted to be a pilot. He was sent to Pre-Flight School in San Antonio, Texas and then on to Primary Flight School in Sikeston, Missouri. His next stop would be to Basic Flight School in Wichita, Kansas where he washed out.

Because he washed out of the flight training, Sheedy was sent to Shepherd Field, Texas where he started his military training. After boot camp, he went to Selfridge Field, Illinois for Radio Operator training. This was followed by three other airfields for more advanced training. He then wound up in Salt Lake City for preparations to go overseas. Sheedy proceeded to Washington State for the debarkation process.

He and many others boarded a ship of the Laurel Line, a commercial line, and began the crossing of the Pacific to Hollandia, New Guinea. They went without escort because the ship could outrun submarines and had the "new" sonar detection devices. They even went into evasive action while crossing before arriving at Hollandia.

He was assigned to the Port Moresby Radio Station. He asked for night duty so he could have time to memorize all the circuits. His only job was to wake up the non-commissioned officer in the event of an emergency. Sheedy learned the systems quickly. He inherited the job of the non-commissioned officer in charge of maintenance and was promoted to Corporal. The Commanding Officer had the power to appoint non-commissioned officers in charge but to get the rating it had to come from headquarters. When they knew it was Sheedy, they promoted him to Sergeant. This earned him a jeep, tent, evening showers, and naps!

While on guard duty at Moresby, many times they had Japanese penetration but it was quickly taken care of.

The evening at the radio tower, Sheedy had 1,000 volts race through his body but survived. Previously a soldier had been killed in a similar incident. An interlock device had failed to operate the door to the shed. The soldier had a knee touch the door as he knelt against the unit and pushed his hands to get away. The current raced through his knee and into his body, killing him instantly.

When the A-Bomb was dropped on Hiroshima, they were getting ready to send personnel from the base to Japan. Sheedy recalls only two times they were called out for formation. The first was the death of President Roosevelt and the second was when the bomb was dropped.

After the war, he then flew to the Philippines and loaded up with others and equipment on LST ships and became part of a convoy. On the way, they became victims of the great typhoon of 1945. Their ship lost control of its rudder and wound up in the center of the convoy for protection. He recalls hearing someone getting so scared and began crying out "prepare to abandon ship!" They did not, luckily, but Sheedy recalls untying his shoes as to not wear them just in case they did abandon ship.

Sheedy was sent to Kobi and Osaka, Japan to set up communications installation and maintenance. He would trouble-shoot to make sure it all worked. Sheedy finally came home and was discharged December 7, 1945. He received the American Service Medal, Asiatic-Pacific Medal, Philippines Liberation Ribbon and World War II Victory Medal.

After the war, he worked in various machine shops and earned a Q-rating to work for civilians making atomic fuel rods. He finished night school at SUNY System College in Farmingdale, New York as a licensed Professional Engineer. After working at LamTec Industry in Long Island, he worked for Koppers Company until his retirement in 1985. He also did some consulting work.

Sheedy then moved to Sykesville in Carroll County, Maryland. He lives with his wife of sixty-six years, Mary Elizabeth Wirtner Sheedy. He has four daughters, nine grandchildren and seven great grandchildren.

Of his wartime experience, Sheedy says he enlisted because he thought it was what he should do. He has the highest respect for what the military gave him. He has depended on those experiences throughout his life.

Conrad Sigmon
Signalman 3rd Class
S.S. George Westinghouse
U.S. Navy Armed Guard
World War II

On the way back from England in a convoy, Signalman 3rd Class Conrad Sigmon admired the calm sea. They then received orders that the Commodore of the convoy decided to have gunnery practice. Any ship that wanted to could fire their weapons. A shell from one of the ships went off course and struck a British ship killing two sailors. Later Sigmon's vessel, the Liberty Ship S.S. George Westinghouse, wound up being close enough to watch the burial at sea of the two British sailors. It was a humbling experience that he still vividly remembers today.

The high school graduate enlisted in the U.S. Navy in April 1944. Born in Rocky Mount, Virginia on November 23, 1925 to Lorton and Blanch Sigmon, the youthful sailor survived the rigors of Boot Camp at Camp Perry in Virginia and was then sent to Bainbridge, Maryland to Signalman school.

While in Signalman school, Sigmon and another trainee were using semaphore flags one evening across campus to send messages to each other complaining about their drill instructor not knowing the instructor was watching them and reading the messages. Needless to say they were both restricted to base. Later, the instructor came to Sigmon and told him about the Navy's Armed Guard, a fairly new concept, on merchant ships. He said it was the best duty in the Navy, but he had to volunteer for it. Sigmon bit.

The sailor was off then to Le Havre, France. He was assigned to the Liberty ship S.S. George Westinghouse. Liberty Ships were cargo ships built in the U.S. during

World War II. They were cheap and quick to build. The quickest ever built was launched in four days, fifteen and a half hours after the keel was laid. Eighteen American shipyards built 2,751 Liberty Ships between 1941 and 1945. They served their purpose even though they were not the prettiest ships in the fleet. President Franklin D. Roosevelt once called them "a dreadful looking object" while Time magazine said they looked like "an ugly duckling." One trip across "the pond" for a Liberty Ship was considered a success.

December 15, 1944 started out as a beautiful day. Fifteen ships were ready to cross the English Channel to get rock ballast and to head back to the States. They were ready to leave when a heavy undertow caused his ship to hit two anchored ships putting a hole in the Westinghouse. The hole was high enough that they could still get underway.

As they crossed the Channel, a storm pounced upon the ships. Winds kicked up and high waves crashed over the vessels. They lost all of their lifeboats and considered going overboard. During the night, they almost collided with twenty other ships. Sigmon believes that this same storm was the one that claimed the life of Big Band leader Glen Miller. His plane disappeared that same night in the same storm.

On the morning of the 16th, the storm had settled down. Calm, glassy seas lay before them but there were no other ships in sight. The Liberty Ship found a buoy near Southampton, had the hole patched up, and rock ballast loaded. They still had one big problem…they could not see any of the other ships.

They finally made contact with the convoy of eighty-six ships but there were supposed to be one hundred and one. The other fifteen had either been heavily damaged or sunk during the storm. Thirty-two Armed Guard perished, not counting their crews. For Sigmon, it had touched him deeply. While getting ready to leave, he had signaled with two other friends he had graduated with from Signalman School. After the devastation from the storm, he never found out if they were dead or alive somewhere.

Two days out, a destroyer escort came alongside and started rolling depth charges to try to destroy a suspected German U-Boat riding underneath the Westinghouse to hide its sonar blip. Two days after that, their steering broke and their ship started making right turns only and came so close to some other ships that they could talk to their crews. The Westinghouse was dead in the water. A destroyer escort stayed behind with them as the convoy moved on. Later, after repairs, they were able to catch up with the other ships.

All the while these events were going on, their ship, like so many other Liberty Ships, began vibrating badly. By then they were close to Boston and hit a monstrous snowstorm. The snow was so bad they could not even see the bow of their own ship. When the storm stopped, they had moved from Row 4 to Row 8 in the convoy and wondered how they had not hit anyone. They, at last, had made it to New York and had plating put on around the ship to stop the vibrations.

Another crossing he would take was to Portuguese South Africa which was pretty uneventful other than Roosevelt had died and the Germans had surrendered! On the way back they secured all watches and lay back on the decks as if it were a cruise liner. According to Sigmon, ratio wise, the Armed Guard lost more men than the Army during World War II. He received many medals and ribbons for his service and was discharged in April 1946.

Sigmon took advantage of the GI Bill and learned carpentry. He worked in the home improvement business for most of his adult life before retiring. The former sailor comes from a family of seven: five brothers of whom one served in the Army, and two sisters. He now lives in Manchester, Maryland since 1992 with his wife of 45 years, Helen Frock, and his two sons and one stepdaughter.

Samuel (Buzz) Smeltzer, PFC, Wireman 87th Infantry Division, 3rd Army World War II

Samuel E., Sr. and Lillian Smeltzer gave birth to their son, Samuel, Jr. on June 27, 1922 in Westminster, Maryland. He finished his schooling at Westminster High School and at the ripe old age of twenty-one, he enlisted in the Army as a Wireman April 4, 1944. His journey would take him to places such as France, Belgium, Germany, and Czechoslovakia.

He was inducted at Ft. Meade, Maryland, went to Ft. Bragg, N.C. in artillery school and changed to communications in artillery. The seventeen-week training class was reduced to six weeks and then to four for infantry training. He arrived at Liverpool, England on a troop transport and went to "Tent City." What amazed him is that they were guarded in a barbed wire camp, surrounded by Italian POW's carrying live ammo in uniforms dyed green.

After arriving in France, he was assigned to the 87th Infantry Division, Rifle Co. B. In Metz, France, he saw his first combat. Afterward, they were loaded on trucks thinking they were going for a rest but headed into Belgium and joined the 3rd Army. The cold weather had taken its toll on his feet. They were nearly frozen. Sam was sent to the rear and found boots on a dead medic that were his size. Those boots took him through the rest of the war.

The young soldier was left behind with others to show the 10th Armored Battalion where the German emplacements were. He also had his first fresh egg and orange since he left home. While passing through the Eiffel Mountains, they fought against pill boxes on the mountainsides. While in

Oberhof, he had the first bath since leaving the states. While bathing, the German citizens told the Germans where to fire their mortars. A shell landed near the room where he was bathing. Sam was not injured.

At the Bulge, they tried to take Malmady but were unable to break through. Patton ordered them to take Malmady whatever the cost. During the battle, while pinned down, a captain he was with ran out, dropped a grenade into a tank two times while a sergeant on a bazooka knocked the track off another and tossed grenades inside. They then relieved the 102nd Paratrooper Division.

Crossing the Mozel River, the division lost wire communications. Sam volunteered to climb a pole, cut wires and attach theirs, allowing them to have communications with command. A sniper began firing at him, just barely missing. Again, he wasn't injured. At 2 AM they were instructed to put their weapons in the bottom of the boat and use their oars and keep rowing until they reached the opposite shore. They hid behind a wall while receiving fire from German 88's, and anti-aircraft fire in two and three waves. The Mozel River lit up like a Christmas tree. They knocked out pillboxes and captured an SS officer who jumped out of a boat and was shot by a soldier. They killed or wounded German soldiers in small towns after crossing the river. They then moved up the Rhine and captured Koblenz, routing the Germans.

Before meeting the Russians, they took up defensive positions near Olsnick. Here is where Sam Smeltzer became somewhat of a hero. He was sent out on a mission in a jeep with a 50 cal. machine gun. As he drove to his destination, he saw Germans on the left, turned around and pointed his machine gun at them, and ordered them to put their hands up. They did! One of them turned out to be a Tank Commander. His troops were surprised to say the least when he had the 45 German POW's march into their camp with their arms raised over their heads. Later, they met up with the Russians in Czechoslovakia.

For his service he received four battle stars, Good Conduct Medal, Infantry Badge, and the European Theatre Medal. Smeltzer's infantry division was the first complete division to come home after the surrender of Germany, leaving Europe on July 4th on the S.S. United States. He was discharged November 11, 1945.

Returning home with hearing loss from the German 88's, he became an employee of the B.C. Herman Plumbing and Heating Business as a plumbing and sales rep, retiring in 1995. Sam now lives in Westminster having been married to Doris Marie Shilling of Finksburg, for sixty-six years before her death in 2007. The old soldier is now retired and enjoys seeing his three sons, the oldest of whom served in the Air Force, and his grand and great grandchildren.

James Robert Thompson
Aviation Radioman 3rd Class
U.S.S. Lexington CV 16
Air Group 94 Bombing Squadron
U.S. Navy
World War II

Pilot Lawrence Haines and Aviation Radioman 3rd Class Bob Thompson were on a bombing mission over the coast of Japan near the Bungo straits in their SB2C Helldiver dive bomber. As they neared the coastline, they began to take flak from antiaircraft fire from the shore batteries. Then they were hit and began to lose oil pressure. They stayed over water and began to descend. Haines glided their craft to the surface of the water and miraculously landed it flat on the ocean. Both Haines and Thompson quickly climbed out, inflated their life raft and climbed in only to watch their bomber succumb to the ocean in less than two minutes. It would be their longest day of the war.

Thompson was born to Andrew and Ethel Mae Thompson on March 7, 1925 in Ednor, Maryland (now Silver Spring). He helped the family on the farm, and it helped him appreciate God's creation. He was captain of the Sherwood High School basketball team and won two championships.

The eighteen-year-old was drafted into the navy on august 25, 1943 and wanted to go to aviation cadet school but didn't pass the test. He attended boot camp at Sampson, New York and then radio school in Memphis, Tennessee, followed by gunnery school in Jacksonville, Florida.

His first flight was on a dive-bombing test at Wildwood, New Jersey. He was told to drink milk and he would not get sick. It didn't help. The pilot did let him take the controls once in a while. During training in Maui, Hawaii he was almost killed. A storm was rolling in and they all were told to return to base. Visibility was bad. As they landed in their plane and came to a stop, they were abruptly slammed in the back by another plane. The propeller missed Thompson by inches. When he was asked if he was okay or wanted to go to Sick Bay, he calmly told the sailors just to get him a roll of toilet paper.

In May 1945, he was assigned to the dive-bombing squadron Air Group 94 on board the U.S.S. Lexington, Aircraft Carrier CV 16 known as the "Blue Ghost." His first mission was a bombing run on Wake Island. Other missions were so cloudy that they were forced to return with bombs unspent. Thompson recalls returning to the carrier with live ammo still on board the aircraft and the very abrupt landings from the arresting cables. He related that the landings were the worse.

On another mission while dropping supplies to POW's, they returned to the Lexington and were almost out of fuel. Aircraft from the Lexington also participated in airstrikes on the Japanese home islands from July through August 1945. Attacks were made on Honshu and Hokkaido airfields, Yokosuka and Kure naval bases and industrial targets in the Tokyo area.

From June 20th to July 18th, 1945, during operations against enemy Japanese forces in the vicinity of Wake Island and the Japanese homeland, he completed his fifth strike against the enemy. He rendered invaluable service to his pilot, Haines, and contributed to the infliction of extensive damage on the enemy.

During an afternoon mission to attack the Kure Naval Base, Haines and Thompson flew his sixth mission with a formation of aircraft that should have taken two and a half hours. During the attack, two planes led in front of them. The enemy shore batteries opened up sending flak and bullets in the air toward the Americans. The Helldiver carrying Haines and Thompson was hit. Their oil line had been ruptured. The pressure to the oil tank had begun to drop rapidly. The pilot led the plane back out over the water as it got lower to the Sea of Japan.

Haines glided the bomber down skirting right over top of the water and the plane came to a halt. The two men began scrambling out of the cockpit, Thompson inflating the raft. The two men jumped into the ocean and climbed into the raft as they watched their plane slowly sink beneath the surface. They were alone. During the attack, two planes were lost.

Thompson and his pilot sat in that raft for eighteen hours in the ocean waiting for rescue that came the next day. He remembers them saying the Rosary and praying. Finally, a seaplane, the Pine Island, came on the scene, rescuing the two sailors, and flew them back to Okinawa. It took them thirty days to get back to the Lexington.

After hostilities, five prisoner of war camps were located. Flights from the Lexington dropped over 1,300 pounds of supplies to them. Members of the crew donated their own shoes, clothing, and toiletries to the air crews who dropped the supplies. The flight crews flew without complaint. The U.S.S. Lexington had the honor of being a part of the formal surrender of Japan in Tokyo Bay.

April 19, 1946, Thompson was honorably discharged receiving the Air Medal for "Meritorious achievement in aerial flight as a rear seat gunner of Air Group 94." The Air Medal is the one which he is most proud of. He also received the Pacific Theatre Ribbon 1-star, American Theatre Ribbon, Victory Medal, and the Air Insignia 3 stars.

Thompson's older brother, Jack, served in the Marines having fought at Guadalcanal, Peleliu, and Okinawa. Both men came home to their families.

On November 24, 1945, he married his high school sweetheart Mary Helen Burroughs. They were married sixty years until her recent death in 2006. With a few jobs behind him, he worked for Cloverly Hardware for thirty years until retirement in 1990. He has three children, five grandchildren and two great grandchildren. The former Helldiver gunner moved to Carroll County in 2003 and now lives in Westminster, Maryland.

He had kept in touch with his former pilot and lifelong friend, Lawrence Haines, until his friend's recent death.

Jack Trenner
Lieutenant
U.S.S. Hodges & U.S.S. Hyman
U.S. Navy
World War II & Korean War

The Destroyer U.S. S. Hyman, DD 732, patrolled Wonsan Harbor, Korea firing at positions called in by U. S. Marines at enemy targets of opportunity. The Officer of the Watch, Lt. Jack Trenner, was relieved by another officer in the Combat Information Center (CIC) at 1555 hours and headed to the ward room for a cup of coffee. At 1600, the enemy shore batteries opened up returning fire at the Hyman.

Shrapnel holes started appearing in the ward room bulkhead from the enemy shells. The Captain sounded General Quarters announcing for all personnel to use inside passageways. Seven minutes later they received a direct hit.

Jack Trenner was born to Frank Trenner and Rose Thiess of Baltimore on July 28, 1925. Jack held a part time job at the Pennsylvania Railroad freight station when he heard of the Japanese attack on Pearl Harbor on December 7, 1941. He graduated from Baltimore Polytechnic Institute (A Course). During that summer he worked for 6-9 months at Bethlehem Steel Shipyard on Key Highway in Baltimore as a Pipe Coverer Helper. He would enter Union College in Schenectady, New York on a scholarship. The S400.00 scholarship would pay his tuition and books for one year. He graduated at 19 with a BS in Electrical Engineering.

Union was a great experience for Trenner. He entered the Navy V-12 program. For four semesters a good portion of the company consisted of men from the fleet. Later, Trenner went on to Columbia University Midshipman School and commission as an Ensign. From '42-'44 he went from a boy to a man and well prepared for boot camp and Midshipman School in the Navy.

After a rigorous month of boot camp at Sampson Naval Training Center on Lake Seneca, New York he left for the Naval Training Center in Miami, Florida. He would wind up going to Jacksonville, Los Angeles, San Francisco and onto Pearl Harbor. Gunnery School waited on Wainai, Hawaii. In July of '45, he sailed on a Jeep Carrier to Leyte Gulf in the Philippines having followed almost to the letter the same path that the infamous U.S.S. Indianapolis had taken when it was torpedoed after delivering the Atomic Bombs to Guam. Meanwhile, the Third, Seventh, and Fifth Fleets were massing for the Invasion of Japan with five million men.

On the night of August 6, 1945 in the Bachelor Officers Quarters as Trenner watched Jack Benny in the movie "The Horn Blows at Midnight," news came that the bomb had been dropped on Hiroshima. The place went crazy. Wild marines broke into the beer cache and they all had a party. All the many ships in the harbor began firing tracers into the air. It looked like the 4th of July.

On August 21, Trenner reported on board the U.S.S. Hodges, a destroyer, as Assistant Chief Engineer. With the second bomb dropped the war was over and a treaty signed. The Hodges was assigned to secure a Japanese weather station on the island of Ita Abba.

In March of 1946, he became Chief Engineering Officer on the Hodges. As a qualified chief engineer, he was in great demand. He would serve on two destroyer escorts before his discharge in June 1946.

In the summer of 1946, he met his future wife, Gloria Peters while he worked at Bethlehem Steel. They were married June 21, 1947. Trenner was required to report to the Naval Reserve Training Center at Fort McHenry in Baltimore, Maryland once a week and a training course for a week once a year. He was in Company 5-5 when the Korean War broke out in the spring of 1950. Trenner was recalled to duty and received orders on November 1 to report to the U.S.S. Hyman in Rhode Island. He had to stop working on the home he was building. His family finished it while he was away.

After six weeks aboard, he was sent to Damage Control School in Philadelphia. From there he went to Edgewood, Maryland for Atomic, Biological and chemical (ABC) School. He met back up with the Hyman in Gulf Juan, France. In October 1951, the Hyman steamed out for the Pacific and Korea. While in route, they lost one seaman who was washed overboard, his body never found.

On November 1951, they refueled from the U.S.S. Tolojawa and entered Wonsan, Korea. They would fire at enemy shore batteries in conjunction with the Marines who occupied islands in the harbor. They would leave the harbor every two or three days for replenishment and refueling. They would also pick up pilots who would have to ditch in the harbor.

On November 24, they began taking fire from the enemy shore batteries. General Quarters sounded and men scrambled to their battle stations. At 1607 hours, they received a direct hit on a reel of mooring line amidships on the starboard side. The line absorbed most of the impact along the steel side-rail. The nose cone of the shell went across the deck and wound up in the coffee bin in the galley. The shell had Russian markings. Luckily no one was injured!

The Hyman would also provide plane guard duty for the carriers of the 7th fleet. One thing that impressed Trenner was awaking in Korea and finding six inches of snow on deck. He thought of the poor soldiers in the foxholes with all of the wind and snow. In October 1952, Trenner was honorably discharged. He was awarded numerous service medals and ribbons including the Combat Action Ribbon.

After the war, Jack went back to Bethlehem Steel Sparrows Point Plant where he worked for thirty five and a half years before retiring for the last twenty eight years.

Since moving to Carroll County in 2003, he now lives at the Fairhaven Retirement Community in Sykesville, Maryland with Gloria, his wife of sixty-three years. They enjoy traveling around the world, having been to numerous countries. Together they enjoy their life and being with their two sons and two grandsons.

Francis N. Utz
Tech Sergeant - B-17's
Top Turret Gunner/Flight Engineer-Army Air Corps
8th Air force, 45th Air Division
96th Bombing Group, 337th Bombing Squadron
World War II

The third mission was brutal. Its destination was a sub repair base in Hamburg, Germany. Flight Engineer and Top Turret Gunner Francis Utz knew the B-17 was taking heavy flak and it was accurate. The craft was getting pummeled. Number 1 and 2 engines were lost and the might fortress began going down. As the pilot feathered the props, the co-pilot froze apparently from fear in his seat. The pilot screamed for Utz to come up and help him straighten out the plane. They began transferring fuel from the damaged left wing. But the plane still continued to drop rapidly from 19,000 feet to 7,000 feet. The earth raced up to meet them.

Francis Utz was born November 18, 1924 in Lineboro, Maryland to Millard and Treva Utz. He was raised there with his 5 brothers and 1 sister. He dropped out of Hampstead High School in his sophomore year to work and help the family. Utz worked at the Baltimore Shipyard building Liberty Ships.

Utz was drafted into the Army Air Corps February 12, 1943, processed at Fort Meade, Maryland and sent to Atlantic City, New Jersey for six weeks of boot camp training. While there, one evening, they awoke to look out through the blackened curtains to see a tanker burning at sea. It had been torpedoed by a German submarine. He then went through three phases of training. The first was at LaGuardia Field in New York at the Academy of Aeronautics. Then off to the Casey Jones School of Aeronautics in Newark for two months. Finally, he went to Roosevelt Field on Long Island where he graduated as an aircraft mechanic. There he was given two choices: aerial gunnery school or crew chief on a bomber. He then arrived at Panama City, Florida at Tyndall Air Base for aerial gunnery school. He vividly remembers the third most difficult obstacle course in the nation at that time. He would also have to take a fifty Caliber machine gun apart and reassemble it ending with doing it blindfolded. They learned aircraft identification and air-to-air A24 Bomber targeting.

Utz then headed to Drew Airfield in Tampa, Florida and was attached to the 3rd Air Force where he received bomber training from January to August 1944. While there he also actually learned to fly the aircraft. The airman also learned the B-17F inside and out and did sub patrols off the Florida coast. From Drew it was off to Langley Field, Virginia and to Dow Field in Bangor, Maine where they picked up a new B-17.

The crew then flew to Goose Bay, Labrador, Iceland, Wales and ended up at Snetterton, England. His first mission was November 10, 1944 to Weisebaden, Germany. He was anxious and wondered how they would fare. Between 10,000-15,000 feet the crew went on oxygen and tested their guns. The flak was light but accurate, penetrating their craft twice. They dropped 38 100 pound bombs. They called the mission a milk run.

The third mission was memorable. They had been hit, lost two engines and were going down. As an 88 shell came through the radio room, they leveled off at 7,000 feet. They finally got away from the flak and target and were pleased to find two P-47 Thunderbolt planes escorting them back, one on each side. The crew threw anything of weight out the windows in an effort to get across the English Channel. The craft skimmed the channel at between 2,000-4000 feet expecting to ditch. They were able to land on an emergency field in England. The co-pilot who had frozen now came around. As they sat down on the runway, the plane spun around. The pilot killed the throttle. The crew got out and looked at the plane. In all, they counted 167 holes. Other squadron members were glad to see them, thinking they had gone down.

On one mission, he saw three craft go down and one had a Major who was on his last mission before going home. On mission 11, the radio operator watched to make sure the bombs dropped and saw one bomb stuck in the bay. Utz scurried back to the bomb bay and carefully, due to the doors being open, kicked the bomb repeatedly with his foot until jarring it loose and releasing it. He remembers how scary it was to look down and see the earth below him through those doors.

Mission 17 over Zeitz, Germany found heavy flak. They lost one ship while bombing an oil refinery. He remembers only seeing 4 chutes open. He vividly remembers waving to a buddy in a nearby plane only to find out he was shot down. On April 17, 1945, they bombed the city of Dresden.

On his 20th mission over Elmshorn, Germany on April 7, 1945, they were to bomb an airfield. They began encountering light flak and enemy aircraft showed up. They fought for 45 minutes and watched dogfights all over the sky. That was the day Utz finally shot an enemy aircraft down. It was at 7 O'clock high. He fired two six round bursts and the craft fell to the earth. They also lost 3 B-17's that day.

In all, Francis Utz flew 27 missions and no one on those missions got a scratch. He received the Air Medal with 3 Oak Leaf Clusters, Gunners Wings, WW II Victory Medal, European Theatre Ribbon with 4 Battle stars, and the Good Conduct Medal. He was discharged October 5, 1945.

After the war, he worked at Wheeler Chevrolet in Hampstead for 30 years as a mechanic. Later he worked Gray and Son, Inc. for 10 years in construction. He finally retired in 1990. He enjoys hunting and fishing. He is also the proud member of the Hampstead Post 200 of the American Legion and a Life Member of the Molleville VFW Post 467 in Westminster, Maryland.

Today, Francis Utz lives in Hampstead, Maryland with his wife of 40 years, Betty I. Reed Dovell Utz. Together they enjoy their 2 stepchildren, 5 step grandchildren and 4 step great grandchildren.

Bradley D. Wentworth
Hospital Corpsman E-3
U.S.N.S. Comfort Hospital Ship
U.S. Navy
Desert Shield/Desert Storm

Saddam Hussein had invaded the country of Kuwait and declared it to be part of Iraq. He wasn't leaving. Tensions in the Gulf teemed with desperation as forces on both sides built up. On board the U.S.N.S. Comfort, a Navy Hospital Ship, Hospital Corpsman Bradley Wentworth was waking up for his shift and getting dressed in the dark. As he dressed in the reddened hue of the lighting in his quarters, he could hear the crackle of the ship's loudspeakers come on. "This is the Captain speaking. We have just gone to war. We have now gone from Operation Desert Shield to Operation Desert Storm." Hospital Corpsman Wentworth knew this was it.

Bradley Wentworth, the son of Alan and Carol Wentworth was born on June 24, 1969. He entered the Delayed Entry Program in the Navy in 1987. However, he did not enter until he graduated from Francis Scott Key High School in June of 1988. The Delayed Entry Program allowed him to negotiate a contract with the Navy for which school he wanted, which coast he was willing to accept, an agreed entry rank, an agreed enlistment term and an agreed entry date. His entry date would be June 19, 1988.

Wentworth took basic training at the Great Lakes Naval Basic Training Command for eight weeks. Following that, he received orders for Hospital Corpsman School at Great Lakes. He graduated in November 1988 and was given assignment aboard the U.S.N.S. Comfort until his discharge in June 1991.

The Comfort is not manned twenty-four hours a day, seven days a week but by a cadre crew of about thirty when not at sea. So, Wentworth reported to the National Naval Medical Center (NNMC) in Bethesda, Maryland under Temporary Assigned Duty (TAD) orders. He arrived in November 1988. While at NNMC, he continued his training and became a Surgical Technologist and assigned to the Operating Room.

In August 1990, the Comfort was activated to the Persian Gulf to support troops being sent there after Iraq invaded Kuwait. The Navy allowed only 300 of a total of 1,200 crew sail to Rota, Spain. The plan was to send additional sailors and officers by aircraft to Rota to meet the Comfort there. This was due to by the Comfort being activated. The NNMC could not allow all 1,200 people to leave at once because command had let its own staff dwindle to dangerous levels. If everyone who was supposed to leave actually left there would not have been enough staff to care for the patients.

Wentworth left Andrews Air Force Base in late August, arrived in Rota, Spain, and reported for duty on the Comfort. There was no place to sleep when they arrived, so they were put up in a base theatre. When almost asleep, a Master Chief came in and instructed the sailors in the last two rows to board a waiting bus outside. He told them there were F-14's in the air and running out of fuel and can't land because of foreign object debris on the runway.

The Comfort set sail from Rota, through the Mediterranean Sea to the Suez Canal (for which he received a citation) and into the Persian Gulf. While in the Gulf, the Comfort would rotate with the U.S.N.S. Mercy. Sometimes when not near the fighting, they would have a "Port Call" and visit places like Bahrain, Abu Dhabi and Dubai. While on one call, the U.S.S. Iwo Jima had injured sailors from a steam leak in a boiler room sent to the Comfort. Wentworth assisted in their initial treatment when they arrived at the

Casualty Receiving. Unfortunately, all the sailors died from third degree burns over 90% of their bodies.

The next morning, Captain Penzine made the unforgettable announcement that they were now at war and should be ready to take on the injured when and if they arrived. Thirty days or so later, the war was over. Even in such a short period the Comfort alone treated 8,000 soldiers/sailors, handled over 700 admissions and performed over 300 surgeries.

Wentworth recalls while giving a presentation while touring Fleet Hospital Five that a soldier referred to him as a "Death Cheater." He was offended at first but when quickly explained to him, he realized that it was a term of respect given to seasoned medics. He felt that he was a good tech before going to the Comfort but when he left he was a great tech because he was exposed to having to think quicker, plan ahead better and do more with less.

The crew received orders that 2/3 of the crew would fly back home. They left the ship on barges and boarded large Deuce and a half's that drove through the streets of Bahrain to the airport. As they rode up the streets, civilians came out and ran alongside them with their arms above their heads, crying for joy and thanking them. The whole thing reminded him of pictures he had seen of World War II, yet he was living it. He will never forget that.

Wentworth reported back to the NNMC where he received the Navy Unit Commendation, Combat Action, Sea Service, National Defense, Southwest Asia with two bronze stars and the Liberation of Kuwait Medals. Brad Wentworth was discharged on July 2, 1991.

Wentworth has continually worked as a Surgical Technologist since 1988 and is currently employed with Med Star (since 1993) as well as being a VSR with the U.S. Dept. of Veterans Affairs, Baltimore Regional Office. He is a Life Member of the Veterans of Foreign Wars 6918 in Harney, Maryland. He stays busy writing a bi-weekly column called

"A Veterans Affair" for the Frederick News Post. He lives in Union Bridge, Maryland with his wife Tracy and their five children: Clarissa, Hannah, Pierce, Nathaniel and Emma.

Arthur Williams
Sergeant Major, U.S. Army
U.S. Army Medical Corps, 96th Infantry Brigade
Combat Medic
Vietnam

The Vietnam War Memorial in Washington, D.C. was one place retired Sergeant Major Art Williams did not want to go to see. He didn't want to go because he felt that when he would look at the names on the wall, he would feel that he did not do his job. Years later, the combat medic who had served bravely in Vietnam finally decided to go see it. When he did, he realized that there were numerous names not on the wall because he DID his job.

Born July 17, 1950 in Baltimore City to Melvin and Athalia Williams, Williams grew into a young man and attended Mergenthaler High School in Southwest Baltimore City. His family knew that all he ever wanted to be was a soldier. Finally, duty called.

Williams served three- and one-half tours in Vietnam from 1968 to 1972. He arrived in Vietnam for his first tour in 1969 after landing in Cam Ranh Bay. His base would be at Duc Pho. He was dropped off in the "middle of nowhere" by himself. Finally, a jeep pulled up with an American soldier inside, was asked his name, and told to get in. Relieved, the young soldier took him to Duc Pho. Williams would serve his first tour as a dog handler. Together, they would sniff our land mines, booby traps, and enemy combatants.

After his first tour, Williams returned home for one month. Then it was back to Vietnam for a second tour as a combat medic with the 96th Infantry. He again started out at

Duc Pho but wound up in Chu Lai for almost nine months. He said he had wanted to be a Physician's Assistant when he left the army but after being a combat medic, he changed his mind.

The soldier noted that while there he saw no color issues. Every man would take care of each other regardless of the color of their skin. Hollywood makes too much of an issue of it. When in war, you are all brothers.

For his third tour, Medic Williams remained a combat medic and saw action at Chu Lai and Da Nang. He became part of the Americal Division that supported the troops in Bam Me Tuit, La Trang, and Da Nang.

Williams' talks about how heavy his rucksack was and some of the equipment he had to carry. The sack with all of his field needs plus medical equipment weighed in excess of fifty to sixty pounds. He also toted a carbine and forty-five automatic. The hardest part of his job was making life and death decisions. Triage was probably the worst part of his job. The young G.I. had to decide who got treatment and who would have to wait.

The worst battle he saw was at LZ Baldie. The battle went back and forth between the American G.I.'s and the Viet Cong (VC). Both sides suffered heavy casualties. Finally, they were able to get out.

Art's last tour lasted for only a half a tour. Returning home, he was met by protestors calling him a baby killer and spat on him at the Chicago Airport. He was finally sent home in 1972 and went to Fort Devens, Massachusetts. He worked at a hospital where he met his first wife. Moves came quickly as he went to Presidio, San Francisco working at a hospital and on to Texas. From Texas, he moved to Fort Dix, New Jersey where he served for two years.

Overseas duty followed as he moved to Germany where his first wife passed away. Once Germany was finished, he was stationed at Fort Meade, Maryland with a return to Texas as Chief Medical Instructor in the Combat Medic

Course. Next came Fort Sheridan, Illinois followed by Fort Leavenworth, Kansas where he made Sergeant Major and remained there for three years. The soldier finalized his duty to his country with retirement in 1992.

War had its consequences on Williams. He became an alcoholic after the war but overcame it. Three years after the war, he received counseling due to the horrific events he saw.

The young soldier received the Bronze star, the purple Heart for shrapnel wounds received, Army Commendation Medal, the Vietnam Cross of Gallantry, Meritorious Service Medal, Legion of Merit, and the Combat Medical Badge.

Williams moved to Carroll County in 1992 after his service in the Army. He went to work for the Chesapeake Health Clinic and also for the Maryland State Board of Medical Examiners. Today he works at McDaniel College in Westminster, Maryland working with the ROTC program. He commends Colonel Schumacher, Commander of the Battalion, for his advancement of the program and says that the kids today are stepping up. They now have the largest freshman class ever. He also is the Assistant State Chief of the Softball Umpires of Maryland Association.

The retired soldier now lives in Eldersburg, Maryland with his wife, Kim, and their two children and one grandchild. Of his service he says that every day was a challenge. If he could change one thing it would be that there not all the names on the Vietnam Memorial Wall. In Vietnam, there was a saying that stated, "For those who have fought for it, freedom has a taste the protected will never know."

Walter Williams, Tech. Sergeant
Cook and Rifleman
2nd Infantry Division,
23rd Infantry Regiment
2nd Battalion, Company F
U.S. Army
World War II

June 8, 1944, D-Day plus 2, the 23rd Infantry Division landed at Omaha Beach, Normandy. Sergeant Walter Williams and his men were staging for a while as the division was getting lined up for an offensive inland. He was assigned to guard duty to watch over a field. Someone called out his name. As they did, an enemy shell hit the bank right near him, covering him with dirt. Several were wounded and three killed. After getting themselves together, he was told to go to the Command Post. The company commander wanted him to go in a jeep and pick up rations. After all, he was a cook. Someone told him if he had not leaned over to find out who called his name, the trajectory of the shell would have cut his head off.

Williams was born January 20, 1919 to Carroll and Laura Williams of what was Oakland Mill in Carroll County, Maryland. He lived there with his brother and four sisters. As a youth, he helped his father on their farm, later working at Oakland Mill in the Melville Woolen Company. As war began to brew in the air, he was drafted into the Army on January 16, 1941, supposedly for one year.

After processing at Fort Meade, Maryland, he went to Fort Sam Houston in Texas for boot camp, later being assigned to the 2nd Infantry Division, 23rd Infantry Regiment. He almost fell on top of a rattlesnake during his training.

Glad to leave there he moved on to Fort Sill, Oklahoma for Artillery School. A new Lieutenant, mean as a snake, was ordered to watch the men sleep one night. The Company Commander did not like him; neither did the men. The next day, Williams had several pies left over. The Lieutenant ordered him to throw them away. Williams said he wanted to save them for the men doing KP. He wound up putting them in the garbage. The Lieutenant would never make it alive through Normandy.

In November 1942, the 2nd moved to Camp McCoy, Wisconsin participating in the first large scale winter training maneuvers ever conducted by the U.S. Army. On skis and snowshoes, the men operated in deep snow at temps as low as 45 degrees below zero.

In September 1943, they arrived in Northern Ireland engaging in intensive training on the bleak Irish Moors preparing for the invasion of France. In April 1944, they arrived in South Wales for final preparations, staging at various British ports. General Dwight D. Eisenhower came to each company and team before they left and talked with them. He said to them, "Some of you will be killed right away, some will be wounded, and some will go all the way through. To those, keep the flag flying because in 60-80 years, America will fall." June 8, 1944, they arrived in Normandy France on Omaha Beach.

They slowly inched their way through the hedgerows, fighting enemy troops and paratroopers. They fought their way through towns such as St. Jean des Baisants, Etouvy, Vire, Truttemer le Grand and Tinchebray. While there, a Chaplain asked him if he was scared, he said yes. The Chaplain led him to the Lord: a day he will never forget. Brest would fall next after 39 days of some of the most savage fighting of the war. They moved across France into Belgium and took up positions along the Siegfried Line. He would be wounded by hostile fire in Belgium in November 1944.

The Battle of the Bulge began on December 16, 1944. Williams recalls it being so cold that their clothes would actually freeze to your backs. The German breakthrough on the 16th faced all three battalions of the 23rd fighting savagely in the line. Beating back tank and infantry attacks upset the German plan to cut off vital supply lines at Liege. General Courtney Hodges, Commanding General of the First Army stated, "What the 2nd Division has done in the last four days will live forever in the history of the U.S. Army."

During the Bulge, he and others loaded supplies in the back of a truck. The driver took off. As they drove, shells began raining down. A shell hit near them, nearly cutting off the foot of a man next to him. Shrapnel came flying through the trucks window, missing Williams' ear.

Christmas came and they were in a captured German Barracks. Williams had several turkeys on gas cookers readying them for a Christmas dinner. A German shell blew off half the building sending debris and plaster over the men and turkeys. After getting more stoves brought up, they cleaned off the turkeys and began cooking again, lucky to be alive. Williams was even able to secure material to make hotcakes for Christmas morning.

At St. Vith, they decimated the 3rd Panzer Division. In their dash to Merseburg, a total of 365 miles in 23 days, they captured thousands of PWs, hundreds of towns and an American PW camp. At the Mulde River, they waited for the Russian Army. While in Pilsen, Czechoslovakia, they were greeted by the population as heroes, a fitting end to a long struggle. Over 30,000 PW's were taken, 13,000 of which were hospitalized, and the 2nd endured 217 continuous days in combat, a record unequalled by no other division.

Along with his 5 Battle Stars, he has received the Good Conduct Medal, the European Theatre of Operations medal, Blue Infantry Combat Badge, and the Purple Heart. Williams was discharged October 27, 1945.

He would return home and work for Beacon Steel for 37 years full time and 5 years part time. In 1946, he married Gloria Boone of New Windsor, Maryland, his wife of 64 years. He has four children and two grandchildren and lived in Westminster, Maryland. Since this writing, he has departed this world.

James C. Woolley
Wire Linesman
Sergeant 1st Class, U.S. Army
8th Infantry Division Pathfinders
World War II

Little did 17-year-old James Woolley Jr. know when he quit high school to enlist in the prewar Army in February 1941 that he would see some of the most vicious combat of World War II.

Woolley was born in Memphis, Tenn. He had two sisters, one of whom died of leukemia when she was three and a brother who graduated from high school with Elvis Presley.

Woolley was sent to Belfast, Northern Ireland, Dec. 5, 1943. On the crossing, the ship encountered rough weather, and the young soldier got seasick along with many others. Training in the months ahead led him to the 8th Infantry Division, also known as the Pathfinders. Their next stop: Normandy, in northern France.

On July 4, 1944, they arrived on Omaha and Utah beaches.

While at Normandy, he was driving a supply truck in a convoy moving forward. While backing up, his truck hit a land mine, destroying the vehicle. From there, the soldiers fought their way through the hedgerows. The 82nd Airborne was replaced on the front by the 8th Infantry Division. Then the 8th fought at Ville de la Haye du Puits, where they took a beating and lost many men. The young switchboard operator was seeing more action than he wanted.

After more fighting, the 8th Infantry arrived in Wiltz, Luxembourg. The small, beautiful town received the troops with open arms.

But now the GIs faced a 20-mile-long front and bitterly cold weather. From September to November, they fought in the Hurtgen Forest. They built foxholes with logs over them to protect themselves from the elements and the enemy.

During the battle, the 1st Army suffered 33,000 casualties. Again, the 8th Infantry Division was switched and took the 28th Infantry Division's place in Hurtgen Forest.

When 28th Infantry was finally relieved, they left the area for Wiltz, only to have the Battle of the Bulge happen there. The 8th Infantry traveled to Le Havre, in northwestern France to ship home after the war and went back through Wiltz. Woolley was amazed to see the once-beautiful town flattened by the battle.

In January 1945, the 8th crossed the Roer River into Duren, Germany. As a forward switchboard operator, Woolley had never seen so much artillery fire. The Germans pounded them.

While near Duren, they found a puppy that took a liking to them. One soldier adopted the dog and named it Roer. Later, the dog would be smuggled on their return ship in a duffel bag and live with the GI for 12 years.

Once they arrived near the Baltic Sea, they eventually met up with the Russians.

One of the true horrors of the war that Woolley witnessed was the liberation of concentration camps. There were a few with POWs and a mix of civilians, but most were civilians of many ethnic groups. He recalls watching small children eating out of garbage cans as they traveled across Europe. Bodies were stacked up like firewood in the camps. The stench was horrible. Woolley said he wanted people to know that the Holocaust was not something made up -- it was very real.

The 8th Infantry Division captured the most prisoners, 250,000, mostly from the last week of the war. Three members of the division were awarded Medals of Honor posthumously.

Woolley was discharged from the Army September 16, 1945, after the war.

Woolley is proud of his service to his country and to the 8th Infantry Division. His home bears many reminders of those days when he was part of the band of brothers. For his service, he was awarded the Good Conduct Medal, American Defense Medal, American Campaign Medal, European Theater Operations Medal, four battle stars, World War Victory Medal, World War II Army of Occupation Medal, National Defense Medal, French Croix de Guerre, Korean War Commemorative Medal, Victory in Europe Commemorative Medal and the U.S. Air Force 50th Anniversary Commemorative Medal.

Woolley joined the Air Force for three years after World War II, just in time for the Korean War, but never had to go there. He then went back to the Army until he retired as Sergeant First Class with ten years' service on June 30, 1952. He worked for Westinghouse for 31 years before retiring. He met Ethel Gipe, of Magnolia, at Edgewood Arsenal and married her in 1951. They had been married for 48 years, prior to her death in 1999. Today, he lives in Finksburg, Maryland with his daughter and son-in-law. Woolley has one son and one daughter, four grandchildren and four great grandchildren.

PFC Richard Yates
K Company, 314th Regiment, 79th Division
U.S. Army
World War II

Young Private First Class (PFC) Richard Yates saw an American tank pull off the side of the road. Tanks were never to sit long in one spot due to enemy fire. The rain near La Haye Du Puits, France was cold. The PFC noticed the exhaust coming from the tank. The soldier got next to the exhaust in hope of getting warm and dried out before the tank would move again. Just then a German spotter saw the tank.

Born on April 23, 1925, young Richard Yates from Baltimore City was lonely. So lonely that he quit High School in his junior year because all of his friends were off to the war. Then he was drafted into the army on August 30, 1943. The eighteen-year-old was sent off to Boot Camp at Camp Gruber in Muskogee, Oklahoma. Upon graduation, they would be replacements for the 42nd (Rainbow Division) Infantry.

Yates was rolled into the 79th Division where in February 1944 they participated in training maneuvers in Arizona and then in Kansas for cold weather training. What amazed him was these were not just young men. The eighteen-year-old was training with "old" men in their thirties. Then in April 1944, they were off to Knutford, England where they trained with live ammunition. Finally, they were off to Southampton as backup for the 90th Division, not expecting to be used. Three days later they were called up.

One week later they debarked on Utah Beach in Normandy. He noted that there were still bodies floating in the surf five days after the initial landing. The 79th went on to Cherbourg, France where Yates saw his first action.

Before going into Cherbourg, Yates recalls an incident he had with a fellow soldier while on guard duty one evening. They were to have one sleep and the other stay awake. The other soldier called out to Yates, awakening him, stating he thought he had heard something. Yates told him to do nothing. Again, the soldier said he heard something. Yates told him to calm down.

Suddenly, the other soldier fired off his entire clip into the darkness. All was still afterward. Yates told him to wait until morning and they would check. At morning light, they saw a stallion staring down at them. Yates laughed telling the other soldier he was glad it wasn't Germans because he had fired all his ammo and entirely missed the horse.

Cherbourg was a deep port needed for the delivery of supplies and manpower after D-Day. On June 26, the 79th captured Fort du Roule which overlooked the city.

With Cherbourg under control, the allies were able to start bringing in heavy equipment and started attacking south and west. Shortly thereafter the 79th attacked a Panzer Division.

The next town was La Haye du Puits, a small town with a railroad they needed to capture to secure supply lines. During the first day of battle, the 314th of the 79th Division, Yates group, approached from the left moving forward toward Bolleville.

The rain didn't help. In fact, it was a cold rain for that time of the year. The American tank pulled over to the side of the road. Yates snuggled up near the exhaust to get warm when a German spotter sighted the tank. The tank moved just in time but not Yates. German 88's fired and a shell landed in the trench where Yates hid, blowing him out fifteen to twenty feet onto the roadway.

Most soldiers taped their pins to their grenades that they carried across their chest so they would not be pulled out accidently. Yates had done this to all but one of his. The concussion of the explosion sucked the air out of the trench, blowing the backpack off Yates. When this happened, the pin was pulled out of the free grenade. He was again blown over to the other side of the road with part of his tibia bone in his leg blown out. Within five seconds, he had been shelled twice and was still alive.

The Battle of La Haye du Puits lasted six days with 2,900 men killed in action. Yates was sent back to England until November 1944. He was then transferred to a hospital at the University of Virginia where he was a patient until he was discharged in March 1945.

Things were dismal after the war. He decided to take advantage of the GI Bill and started schooling in 1947. He worked for the Army as a civilian in the Inspector General's Office until 1986.

After moving to Carroll County, Maryland in 1963, he decided to get involved in politics. He was elected to the Carroll County Board of Commissioners and held office from 1994 to 1998. He even got his pilot's license but does not fly anymore.

Since the war, he has been back to La Haye du Puits four times to try to find the exact spot where he met his demise but is still unable to find it. He has become friends with a French family who puts him and his family up when he visits. One thing they tell him each time is that Americans don't understand what it is like to lose their freedom. When the Germans took over their country, they became their slaves; the Germans were their Masters.

Of the medals he received, including the Purple Heart, the one he is most proud of is his Combat Infantry Medal.

Today, Richard Yates lives with his family in Eldersburg, Maryland. He is married to his wife of fifty one years, Mary Yates. They enjoy their family of one girl and two boys and their eight grandchildren. Of the war, he says that it was kill or be killed. You just did what you had to do just to stay alive.

Dennis Yingling, Lt. Colonel 751st (m) Tank Battalion GHQ World War II

Dennis Yingling was born on a family farm in Pleasant Valley located in Carroll County. The farm belonged to his father, C. Tobias and his mother, Ruth Henrietta. He had three brothers and one sister. Dennis helped to work the farm with his family until he attended Western Maryland College where he graduated in 1935, having also taken ROTC. Upon graduation, he worked for Commercial Credit Corporation in Baltimore, traveling around the country on different assignments.

In February 1941, he exercised his commission as 1st Lieutenant and was assigned to the 2nd Armored Division after completing training with the 66th Infantry Light Tanks at Ft. Meade. He states that tanks were in their infancy and there were not many of them before WW II. Because of this, during training, they would sometimes pretend Dodge pickup trucks were tanks.

Dennis was shipped overseas to England in August 1942 until January 1943 when he was assigned to Oran in North Africa with the 34th Infantry Division until the end of the African campaign.

Mr. Yingling was involved in two of the bigger battles of WW II. The first was Salerno, Italy in September of 1943. He was with the 2nd Corps troops assigned with the task force of the 36th Infantry Division. After the landing, they were folded into the 3rd Infantry Division.

The other major battle was more famous: Anzio. In February 1944, he participated in an amphibious landing at Anzio assigned to a task force of the 3rd Infantry Division.

Anzio was the hardest. The allied troops could not move during daylight hours because the Germans had the high ground in the mountains, within firing range of the G.I.'s. Any movement had to be at night. His tanks provided artillery support for the troops. He remembers because of their vulnerability that they had to have evacuation plans ready to go.

He became Captain of a service company and in October 1944, he was promoted to Major and sent to Ft. Leavenworth, Kansas for command and general staff school. Dennis graduated in January 1945 and made Executive Officer of his battalion. Upon completion, he was sent back to Italy just north of Pisa. He saw the end of the war in Europe from Milan, Italy. After the war, he continued in the Reserves.

During the Korean War, Public Law 49 required the pulling up of reserve units. As such, Dennis was promoted to Lt. Colonel, eventually completing his active reserve duty.

For his service, some of the medals he received were the Bronze Star, National Defense Service Medal, European African Middle Eastern Campaign Medal, American Campaign Medal, Armed Forces Reserve Medal, Good Conduct Medal, and Amphibious Landing Medal.

When he completed his service, he returned to work at Commercial Credit Corporation. In 1946, he was sent to Roanoke, Virginia to open an office there. In 1955, he resigned from Commercial Credit and worked for Carroll County Bank and Trust doing consumer loans. He eventually retired in 1982 as the President.

Dennis has three brothers and one sister. His younger brother was assigned to the 110th Field Artillery during WW II. Another brother worked as a civilian for the U.S. Navy. Since the initial writing of his story, Mr. Yingling has departed this world.

The life he had led has been remarkable. His service to his country was equally outstanding. During his service, he saw things that we all hope none of us will ever have to see in our lifetimes. It is with a thankful and grateful heart that we all should remember Dennis with admiration.

About the Author

Bill Hall was born in Baltimore, Maryland in 1948. As a child, he always wrote short stories and even sent one to President Eisenhower and got a response from him at age 11. As he grew older, duty called during the Vietnam conflict and he enlisted in the U.S. Coast Guard. After four years, he was discharged and became a firefighter in the Baltimore City Fire Department, serving for 30 years.

After hearing so many people tell him he should write a book about his experiences, he decided he would. **Turnout-A Firefighters Story** was his first true venture into the field of published writing. It was a success! He has since written and published a book about his hometown: **Sykesville**, Maryland. His latest book is his first published fiction. It is a Christian novel titled, "**Remember Me in Paradise-The Story of the Thief**." He has several other projects completed and in the works that he hopes to get published soon. Currently, he has finished writing a weekly newspaper series on veterans which appeared for thirteen months.

He is the father of five wonderful daughters: Amy, Emily, Holly, Bethany, and Ashley and has five grandchildren. He is married to Carol Beth Hughes Hall of Sykesville and they are Born-Again Christians. After retiring from thirty years in the fire department, he spent a year as Director of Fire and Safety at Springfield Hospital Center in Sykesville before going to work for Carroll County Government as an Emergency Management Planner and then Director in Homeland Security. He currently works for Northrop Grumman Corporation. In his "spare" time he has written several stories, fiction and non-fiction, which are looking for a home at a publishing company.